Electrical Engineering License Exam File

Sixth Edition

Lincoln D. Jones, P.E.
San Jose State University

ENGINEERING PRESS, INC. SAN JOSE, CALIFORNIA

The first five editions were titled
ELECTRICAL ENGINEERING LICENSE REVIEW

Donald G. Newnan, Ph.D.
Exam File Series Editor

Printed in the United States of America

5 4 3 2 1

Library of Congress Cataloging-in-Publication Data

Jones. Lincoln D.
Electrical engineering license exam file / Lincoln D. Jones. -- 6th ed.
p. cm. -- (Exam file series)
Rev. ed. of: Electrical engineering license review.
ISBN 0-910554-71-4 :
1. Electric engineering -- Examinations, questions, etc.
2. Electric engineering -- Problems, exercises, etc. I. Jones, Lincoln D. Electrical engineering license review. II. Title.
III. Series.
TK169.J66 1989 88-31943
621.319'24'076--dc19 CIP

Engineering Press, Inc. P.O. Box 1 San Jose, California 95103-0001

Contents

Preface

This edition reflects the author's evaluation of the kinds and distribution of problems that have appeared in past National Electrical Engineering Exams (also called the Principles and Practice of Engineering Exam) prepared by the National Council of Engineering Examiners (NCEE). The examination is described in detail in the Introduction.

Many changes have been made because of the changes in the problem categories made by NCEE in 1986 and 1988. Most of these changes are reflected in chapters 7 and 8. Because of the addition of a multiple choice problem format, some new problems have been added to conform to these new guidelines. Although NCEE combined problems in the machinery and power distribution area all into a power heading, I concluded that this did not justify a change in this edition. While the area of communications (and electro-magnetic fields) no longer has a special heading, but is spread out through the other areas, it was decided to keep these chapters as they are. The area that has definitely been eliminated by NCEE, is that of illumination; that chapter has been deleted and replaced with computers. Because the NCEE has not allowed the publication of their exam problems, none of them are reproduced in this book.

Every effort has been made to give the proper and correct solutions. There may be errors in the book, partly due to different possible interpretations. Any corrections or errors found would be gratefully received. Many people have aided and contributed to the preparation of the book I am particularly grateful to Professors Jack Peterson (Santa Clara University), Michael O'Flynn, Donald Newnan, and Rengaiya Rao (San Jose State University) and to Howard Plotkin (Pennzoil Company).

Lincoln D. Jones

1

Introduction

This book is for electrical engineers seeking professional registration. The vast majority of engineers obtain registration by passing, in turn, two national examinations.

The first examination is the 8-hour Engineering Fundamentals exam. The National Council of Engineering Examiners (NCEE), which prepares and grades the examination, calls it the *Fundamentals of Engineering* exam. The various state engineering boards call it by a variety of names, including *Engineer-In-Training, E.I.T.* and *Intern Engineer.* Engineers in all branches of engineering take the same examination. Review materials for this initial examination are the subject of other books and are not included here.

The second of the two national engineering exams is called *PRINCIPLES AND PRACTICE OF ENGINEERING* by NCEE, and is usually referred to as the *Professional Engineer* or "The *P.E.* exam". At the present time it is used by 55 states, districts or jurisdictions (i.e., District of Columbia, Guam, Puerto Rico, Virgin Islands, and North Marianas Islands). Most states (33) give the exam by discipline and the others give it as a combined examination (i.e., choice of questions may be from any of the disciplines). Of course, for your particular area, you should check with your state or district. There is a *PRINCIPLES AND PRACTICE EXAMINATION* written for each of the major branches of engineering. The subject material of this book is for electrical engineers.

Several examination formats have been used. At one time, 20 problems to be solved were presented (with 10 being in the first 4-hour morning session and the others in the 4-hour afternoon session); the examinee then selected four problems in the morning and four in the afternoon. These were graded problems, each problem worth 10 points so that a total of 80 points for the day would be possible; a passing grade was, and still is, 48 out of 80 points. The approximate distribution of problems was as follows.:

Power Systems	3
Machines	2
Electronics	2
Communications	2
Circuits	3
Controls	2
Economics	2
Instrumentation	2
Illumination	2
	20 Problems

This list reflected the way NCEE characterized the problems in the past and in one sense still does; however in 1986, the number of questions was increased from 20 to 24, and, from a "job analysis survey" for electrical engineers, the 1986 breakdown then changed to:

Power	8 (4 in a.m. and 4 in p.m.)
Electronics	6
Controls	5
Logic & Computers	4
Economics	1
	24 Problems

These, again, were graded problems that were referred to as essay type questions.

Starting in 1988, some problems became multiple choice and the rest remained essay type. The multiple choice problems each have 10 parts, with each part offering a selection of five possible answers. This revised format still requires the solution of eight 10-point problems (four in the morning and four in the afternoon). In the morning, and again in the afternoon, three multiple choice problems have been included in the following groupings: Circuits/Electronics, Control Systems/Computer, and Power. Also, the Engineering Economics problem is multiple choice.

Please note that the subject of illumination has been eliminated while machinery, communications, and communications areas are not listed as separate subjects but will be spread throughout the other subjects (the economics area is retained as it is common to all disciplines). And, there will be somewhat more emphasis in the control systems area and that of digital design, computers and programming, simulation, and so on.

Effective with the 1986 examination administration, the technical specality area for electrical engineering encompassed the following problem material guidelines:

1. POWER: Eight problems relating to power transmission lines; power circuits and systems (models, analysis, voltage regulation, load flow, operation, short circuit calculations - balanced and unbalanced); energy management and conservation; power factor correction; plant and building power distribution system; substation design; circuit breaker ANSII standard rating structure and applications; transformer test, application and connections (phasing diagrams) motors (DC and AC induction and synchronous, effects of voltage and frequency variations, speed-torque relations, speed control systems including variable frequency power systems and motor starting); synchronous and induction generators; batteries, National Electrical Code and other applicable codes and standards.

2. ELECTRONICS: Six problems relating to bipolar transistors; FET's; integrated circuits; linear and nonlinear amplifiers; pulse and synchronizing circuits; operational amplifiers and applications; modulation and detection circuits; antennas; communication transmission lines; electrical and electronic instruments; filters; transducers and sensors, instrumentation systems including biomedical applications and interfacing.

3. CONTROL SYSTEMS: Five problems relating to continuous and discrete control systems; bode plots; phase and gain; margins; root loci; conditional stability; effects of noise on system response; interaction with mechanical, electrical and biological systems; use of sophisticated control schemes to effect safe, efficient operation of equipment, processes, and so on.

4. COMPUTERS: Four problems relating to digital hardware analysis and design (combinational and sequential); integrated circuit digital devices (counters, multiplexers, ROM, PLA, etc.); microprocessors and applications; programming (in higher level language, such

as BASIC, FORTRAN, or PASCAL) for dynamic analysis and design; digital computer simulation (such as CSMP); network planning; data communications.

5. ENGINEERING ECONOMICS: One problem relating to engineering and common to all disciplines.

-From *NCEE* (8610) Examination Administration.

Because of the changing quality of the examination, this review book has also changed in appropriate areas. This edition reflects these changes by strengthening the area of control systems, and adding a chapter in the area of digital logic, digital design, computers, programming and simulation. In addition to the many problems and their solutions (in the form of essay answers - i.e., worked out solutions with the reasoning behind the method of solution), the author has included several multiple choice problems to give the "flavor" of the new examination format. Although all the exam questions should be solvable in less than an hour (4 to 6 minutes for each part), *the multiple choice ones presented in this book may be somewhat longer and more detailed than might be expected on the actual exam.*

The authors of the actual questions are asked to prepare their questions such that the examinee may quickly discern whether or not the problem is within his or her competence (considering the time constraints). The concept of the entire examination is that of attempting to determine the minimum competency of the examinee in his or her chosen field(s) of engineering practice; this is in contrast to a typical college examination where the competency level is broken down to grades. The examination is not necessarily designed for those who may have chosen to take graduate work, but, rather those who have developed a reasonable expertise in several subject areas and can demonstrate minimal competency in their professional practice.

As in any examination, one wants the maximum possible points. The examiner that assigns these points must make his judgment based on what is written down. Partial credit is given for partially correct essay type solutions. Therefore, in addition to reasonable neatness, state any assumptions that you consider necessary to allow you to work the problem properly, and provide sufficient explanation so the examiner can judge your reasoning. Assumptions should, of course, follow the logic and requirements of the problem.

Unfortunately, many electrical engineers find that the scope of the *Principles and Practice of Engineering* exam is somewhat broader than their own day-to-day activities. As a result it is desirable for most engineers to undertake a substantial technical review program before attempting the exam. This book has been prepared to help organize that review effort.

2

Basic Circuit Analysis

In reviewing dozens of examination problems in this category, certain types of problems appear frequently. They are:

- Wave form analysis
- Transient analysis
- Resonance
- Meters & form factors
- Bridge circuits
- Impedance matching
- Bandwidth calculations

Transient analysis usually involves dc steady state conditions and then a switch suddenly opening or closing. Other types of problems were A.C. steady state types which could be solved by phasor methods.

The meter problems involved three basic types.

1. Shunts and series resistors to expand the meter's range.
2. True Rms vs average reading meters and various non-sinusoidal inputs.
3. Meters in bridge circuits.

A guide to solving transient problems involving d.c. sources can be developed. To find the steady state conditions prior to the opening or closing of a switch, imagine all capacitors to be open circuits and all inductors to be short circuits. Compute the voltages across these open circuits and the currents thru these short circuits. The values found this way can be used as initial conditions for the circuit at the 0+ time (immediately after switch is changed). Exceptions to this rule are found when an impulse of voltage or current occurs. An example would be connecting two capacitors with different voltage-charges together in parallel with no series resistance between them. This kind of problem can be handled by a "conservation of charge" principle. That is, the total charge on the two capacitors must be the same before and after the connection. A similar condition is true for two series inductors with no shunt resistances. In this case, a "conservation of flux linkages" must hold.

For the transient solution, systems involving a single loop or node analysis are generally solved most easily by writing a linear differential equation and solving by classical methods. If, however, the system involves several loops or nodes, a system of equations using the Laplace variable 's' should be written and solved by matrix methods. Recognizing that the solution will be of the form $k_1 \epsilon^{s_1 t}$, a term for each value of s found from the characteristic equation (the denominator polynomial in s) found from the matrix solution of the network equations.

Impedance matching problems fall into three types. The simplest involve the use of a transformer and its turns ratio to make the load impedance appear to equal the source impedance. Problems of this type usually look at a loudspeaker being connected to some kind of an amplifier.

A second and more interesting problem is that of adding some complex impedance in parallel with a load resistance so that the combined parallel impedance is now the complex conjugate of the source impedance.

A third type of impedance matching problem is one where the real parts of both the source and load impedance cannot be altered but a series reactance with the load can cancel the reactance of the source so that the maximum amount of current under these conditions can flow, thus maximizing the power in the load.

Meter problems involving form factors such as a peak reading meter being calibrated to read the RMS value of a sine wave. This would be equivalent to a form factor of 0.707. Thus the meter would read 70.7% of the peak value of the wave form applied. Then, for example, if a triangular wave were applied, a correction would be necessary to find the true RMS value. Take, for example, a triangular wave of 10 volts peak which would have an Rms value of 5.77 but would read 7.07 on such a meter. Thus a correction factor of 5.77/7.07 = 0.82 must be applied to each such reading on this meter for the true RMS value of a triangular wave. The RMS value of a non-sinusoidal wave can be found by integrating the square of the function over a period, dividing the result by the period to find the average square, and then taking its square-root.

Finally, the meter shunt problem can usually be simplified by assuming the current thru the shunt is much larger than that thru the meter. When current of 10 ma or larger are being measured on a meter whose full scale deflection requires 50 μa, the error caused by this assumption is about .5%. As the measured currents become larger, the error is smaller.

With this assumption then all that is necessary is to know the meter movement's internal resistance plus any series resistance placed in the meter circuit. One may then calculate the voltage drop across the meter and series resistance for the required full scale deflection. Then this voltage divided by the intended current to be measured gives the value in ohms for the necessary shunt.

Examples of these various problems and their solutions follow.

CIRCUITS 1

A 50 micro-ampere meter movement with a 2 kilohm internal resistance is to be used with an Ayrton shunt arrangment so as to have full scale ranges of 10 ma, 100 ma, 500 ma, and 10 amps. The maximum voltage drop across the input terminals for a full scale deflection should not exceed 250 millivolts.

Determine the resistor sizes needed and state any assumptions made.

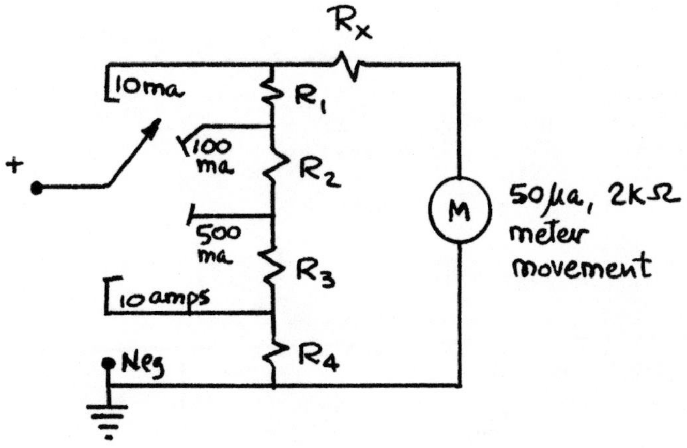

SOLUTION

R_X can be determined from the voltage limit. At 0.250 volts, the current through the meter must be 50μa when using the 10ma setting.

Then $\frac{0.25}{R_x + 2K} = 50 \times 10^{-6}$, $\quad R_x = 3K\Omega$

Now the current through the shunt string of $R_1 + R_2 + R_3 + R_4$ must be 10ma-50μa when the voltage is 0.250 volts.

$10 \times 10^{-3} - 50 \times 10^{-6} = 9.95$ ma

This is only 0.5% less than 10 ma

The accuracy of our computations will not be seriously affected by ignoring this small amount when calculating the value of $R_1 + R_2 + R_3 + R_4$.

$$\text{Then } R_1 + R_2 + R_3 + R_4 = \frac{0.25}{10\text{ma}} = 25\,\Omega$$

Now with the switch in 100 ma position, the equivalent circuit looks like this:

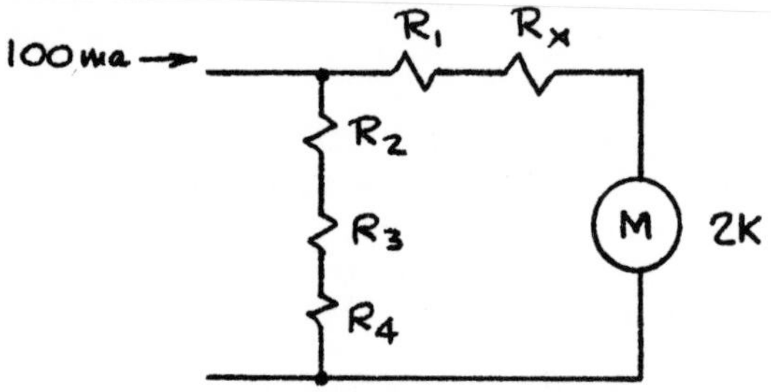

Assume $R_1 \ll R_x + 2K$ since $R_1 + R_2 + R_3 + R_4 = 25$

and $R_x + 2K = 5K$ this is at worst a 0.5% error if $R_2 + R_3 + R_4 = 0$.

Then $$R_2 + R_3 + R_4 \cong \frac{0.25}{100\times10^{-3}} = 2.5\,\Omega$$

So $R_1 = 25 - 2.5 = 22.5\,\Omega$

Now with the switch in 500 ma position:

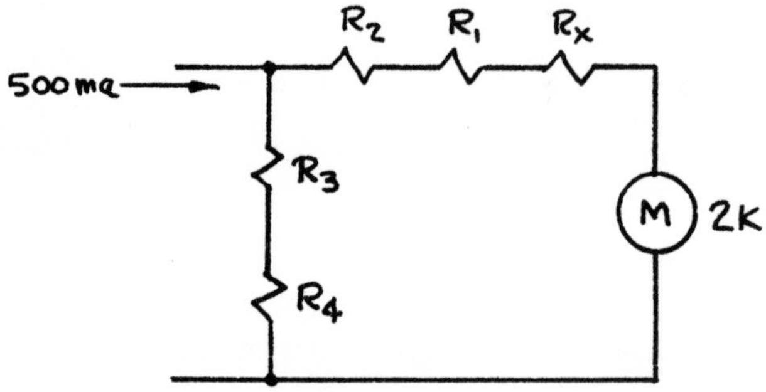

Again, the error of neglecting $R_2 + R_1 \ll R_x + 2K$

is less than 0.5% so

$$R_3 + R_4 \cong \frac{0.25}{500\times10^{-3}} = 0.5\,\Omega$$

So $R_2 = 2.5 - 0.5 = 2\,\Omega$

Finally, the switch at the position for 10 amps:

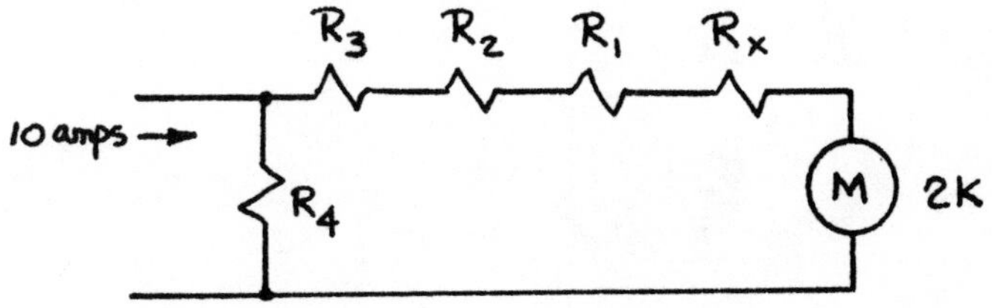

As before, $R_3 + R_2 + R_1 \ll R_x + 2K$

So $R_4 \cong \frac{0.25}{10} = 0.025\,\Omega$

and $R_3 = 0.5 - 0.025 = 0.475\,\Omega$

The absolute value of these resistors could be calculated without these simplifying assumptions by writing simultaneous equations for both parallel branches of the circuit in each case. Since resistors are seldom available in better than 1% tolerances, or at best 1/2%, the added labor for this additional accuracy is not warranted.

CIRCUITS 2

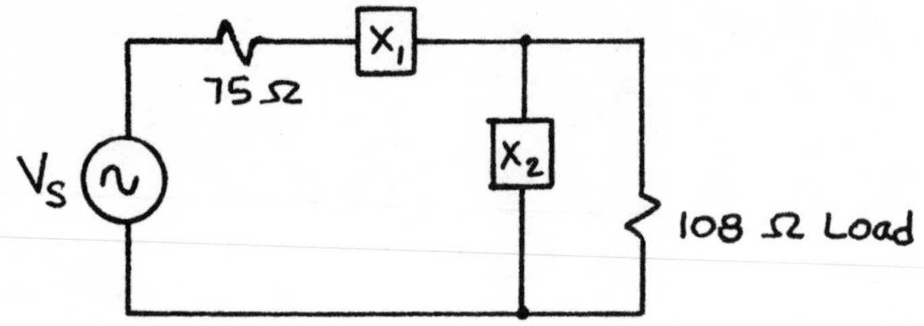

V_s is a 60 hz source. It is desired to maximize the power delivered to the 108Ω resistor. The 75Ω resistance is internal to the source. X_1 and X_2 are reactive elements (capacitors or inductors) to be added. Find appropriate values for these elements so that the power to the load is maximum.

SOLUTION

Assume that X_1 is part of the source impedance and X_2 will be part of the load impedance. Then maximum power will be delivered to the load when

$$Z_S = Z_S^*$$ (The load is the complex conjugate of the source impedance.)

Assume X_1 to be inductive so

$$Z_S = 75 + jX_1$$

Then the parallel load impedance is capacitive,

$$\frac{108(-jX_2)}{108 - jX_2} = \frac{(108)\left(\frac{1}{j\omega C}\right)}{108 + \frac{1}{j\omega C}} \left(\frac{108 - \frac{1}{j\omega C}}{108 - \frac{1}{j\omega C}}\right)$$

$$= \frac{108}{1 + \omega^2 (72)^2 C^2} - \frac{j\omega (108)^2 C}{1 + \omega^2 (108)^2 C^2}$$

For a conjugate match, the real part of the load must be equal to the imaginary part, thus $75 = \dfrac{108}{1 + \omega^2 (108)^2 C^2}$

$$\omega = 2\pi 60 = 377$$

$$75(1 + 1.66 \times 10^9 C^2) = 108$$

$$C^2 = \frac{0.440}{1.66 \times 10^9} = 265 \times 10^{-12}$$

$$C = 16.3 \ \mu f$$

Then for complete matching, the reactive parts must cancel each other.

For the load we have $\dfrac{-j(\omega)(100)^2 C}{1 + \omega^2 (108)^2 C^2} = \dfrac{-j\,71.7}{1 + 0.439} = -j49.8$

Then $|jX_1| = |-j49.8|$

$$\omega L = 49.8 \qquad Z = \overset{*}{Z}_L = (75 + j49.8)\ \Omega$$

$$L = \frac{49.8}{377} = 0.132 \text{ henrys}$$

An equally valid solution can be found by assuming the load to be inductive and the source capacitive.

CIRCUITS 3

An amplifier of voltage gain equal to 70 dB is to be built by cascading transistor stages that have a gain-bandwidth product of 3 x 10^8 radians/sec. Find the maximum bandwidth that can be achieved using these stages, and the total number of stages required.

SOLUTION

A voltage gain of 70 dB,

$$20 \log v_o/v_i = 70 \; ; \quad \log v_o/v_i = 3.5$$

$$v_o/v_i = 3162$$

For optimum bandwidth, the individual stage gain should be 1.65 (see *Modern Electronic Circuit Design* by David Comer, Addison-Wesley, 1976).

Then $1.65^n = 3162$

$$n \log 1.65 = \log 3162$$

$$0.22n = 3.5$$

$$n = 16.09$$

Since an integral number of stages is required, we select 17. This leads to an individual stage bandwidth of

$$BW = \frac{3 \times 10^8}{1.65} = 182 \times 10^6 \text{ Rad/sec}$$

Bandwidth shrinkage due to cascading:

$$BW_{overall} = (\text{Bandwidth of single stage})\left(\sqrt{2^{\frac{1}{n}} - 1}\right)$$

$$= 182 \times 10^6 \left(\sqrt{2^{\frac{1}{17}} - 1}\right)$$

$$BW = 37.1 \times 10^6 \text{ Rad/Sec} \rightarrow \frac{37.1 \times 10^6}{2\pi} = 5.91 \times 10^6 \text{ hz}$$

CIRCUITS 4

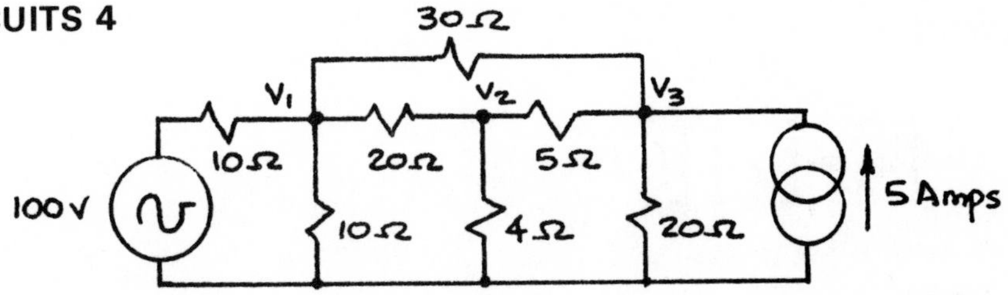

Determine the node voltages, V_1, V_2, and V_3 by setting up the proper nodal equations in matrix form, manipulating the matrices to form the matrix solution for the voltages.

SOLUTION

To simplify the equations we may first replace the voltage source with its series resistance with a Norton equivalent curcuit.

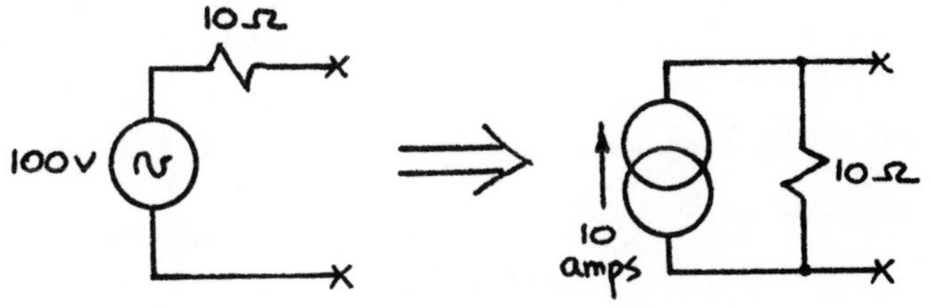

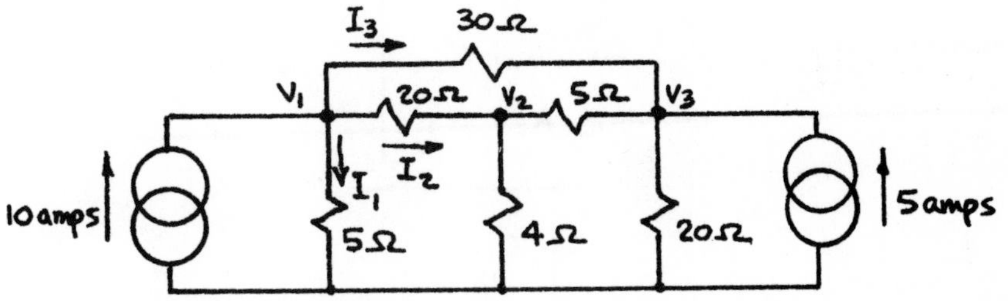

Note that the 10Ω resistor from the source is now in parallel with the 10Ω resistor of the network, making it effectively a 5Ω parallel branch.

The nodal equations are:

$$10 = V_1\left(\frac{1}{5}+\frac{1}{20}+\frac{1}{30}\right) - V_2\left(\frac{1}{20}\right) - V_3\left(\frac{1}{30}\right)$$

$$0 = -V_1\left(\frac{1}{20}\right) + V_2\left(\frac{1}{20}+\frac{1}{4}+\frac{1}{5}\right) - V_3\left(\frac{1}{5}\right)$$

$$5 = -V_1\left(\frac{1}{30}\right) - V_2\left(\frac{1}{5}\right) + V_3\left(\frac{1}{20}+\frac{1}{5}+\frac{1}{30}\right)$$

This is in matrix form

$$[I] = [Y][V]$$

Where

$$\begin{bmatrix} 10 \\ 0 \\ 5 \end{bmatrix} = \begin{bmatrix} 0.2833 & -0.05 & -0.0333 \\ -0.05 & +0.5 & -0.2 \\ -0.0333 & -0.2 & 0.2833 \end{bmatrix} \begin{bmatrix} V_1 \\ V_2 \\ V_3 \end{bmatrix}$$

The matrix solution is

$$[V] = [Y^{-1}][I]$$

$$\text{where } [Y^{-1}] = \frac{\begin{bmatrix} A_{11} & A_{21} & A_{31} \\ A_{12} & A_{22} & A_{32} \\ A_{13} & A_{23} & A_{33} \end{bmatrix}}{|Y|}$$

Where A_{ij} = Signed minor of Y_{ij} (cofactor)

(Reference: Lipshutz: *Theory and Problems of Linear Algebra*. Schaum's Outline Series, McGraw-Hill, Chapter 8.)

and $|Y|$ is the determinant of the Y matrix.

$$|Y| = 0.0269$$

$$\text{So } Y^{-1} = \frac{\begin{bmatrix} +\begin{vmatrix} .5 & -.2 \\ -.2 & .2833 \end{vmatrix} & -\begin{vmatrix} -.05 & -.0333 \\ -.2 & .2833 \end{vmatrix} & +\begin{vmatrix} -.05 & -.0333 \\ .5 & -.2 \end{vmatrix} \\ -\begin{vmatrix} -.05 & -.2 \\ -.0333 & .2833 \end{vmatrix} & +\begin{vmatrix} .2833 & -.0333 \\ -.0333 & .2833 \end{vmatrix} & -\begin{vmatrix} .2833 & -.0333 \\ -.05 & -.2 \end{vmatrix} \\ +\begin{vmatrix} -.05 & .5 \\ -.0333 & -.2 \end{vmatrix} & -\begin{vmatrix} .2833 & -.05 \\ -.0333 & -.2 \end{vmatrix} & +\begin{vmatrix} .2833 & -.05 \\ -.05 & .5 \end{vmatrix} \end{bmatrix}}{0.0269}$$

$$Y^{-1} = \begin{bmatrix} 3.7807 & .7732 & .9926 \\ .7732 & 2.9442 & 2.1673 \\ .9926 & 2.1673 & 5.1747 \end{bmatrix}$$

$$\begin{bmatrix} V_1 \\ V_2 \\ V_3 \end{bmatrix} = \begin{bmatrix} & & \\ & Y^{-1} & \\ & & \end{bmatrix} \begin{bmatrix} 10 \\ 0 \\ 5 \end{bmatrix}$$

Multiplying row by column

$$V_1 = 10(3.7807) + 0(0.7732) + 5(0.9926) = 42.77 \text{ volts}$$

$$V_2 = 10(0.7732) + 0(2.9442) + 5(2.1673) = 18.57 \text{ volts}$$

$$V_3 = 10(0.9926) + 0(2.1673) + 5(5.1747) = 35.80 \text{ volts}$$

Check node no. 1

$$I_1 = \frac{42.8}{5} = 8.56 \text{ Amps} \qquad I_2 = \frac{42.8 - 18.57}{20} = 1.21 \text{ Amps}$$

$$I_3 = \frac{42.8 - 35.8}{30} = 0.23 \text{ Amps}$$

$I_1 + I_2 + I_3 = 8.56 + 1.21 + 0.23 = 10$ which satisfies the original equation.

CIRCUITS 5

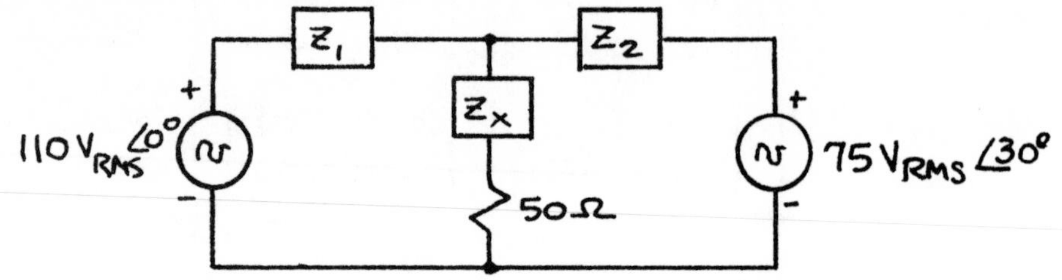

f = 60 hz Z_1 = 60 + j80 Ω Z_2 = 30 - j50 Ω

The 50Ω load is to receive maximum power from this system. A small series reactance, Z_x is to be placed in series with the load to accomplish this. Find the proper element for Z_x and calculate the power into the 50Ω load when this element is placed in the circuit.

SOLUTION

This problem may be simplified by first finding a Thevenin equivalent circuit for the two sources.

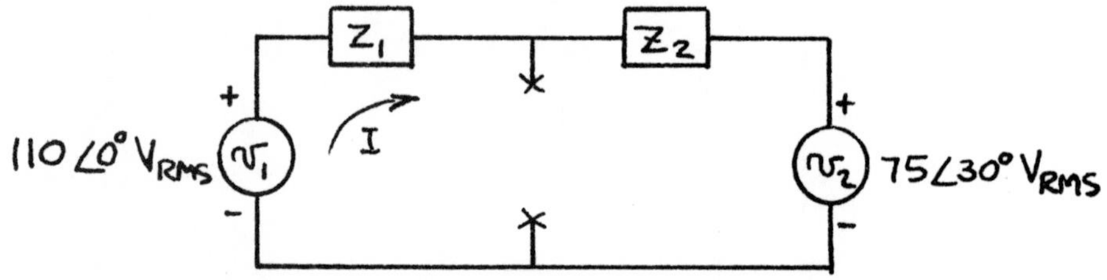

$V_{xx} = V_1 - I_{Z_1}$ and $I = \dfrac{V_1 - V_2}{Z_1 + Z_2}$

$$I = \frac{110\angle 0° - 75\angle 30°}{60 + j80 + 30 - j50} = \frac{110 + j0 - (64.95 + j37.5)}{90 + j30}$$

$$I = \frac{58.61\angle -39.78°}{94.87\angle 18.43°} = 0.62\angle -58.2° \text{ Amps Rms}$$

$$\text{then } V_{xx} = 110\angle 0^\circ - 0.62\angle -58.2^\circ (60 + j80)$$
$$= 110\angle 0^\circ - (0.62\angle -58.2^\circ)(100\angle 53.1^\circ)$$
$$= 110\angle 0^\circ - 62\angle -6.1^\circ$$
$$= 48.8\angle -7.7^\circ \ V_{Rms} = V_{Th}$$

To find the Thevenin impedance, we look into terminals *XX* with all sources disabled.

$$Z_{Th} = \frac{Z_1 Z_2}{Z_1 + Z_2} = \frac{(60 + j80)(30 - j50)}{90 + j30}$$
$$= \frac{(100\angle 53.1^\circ)(58.3\angle -59^\circ)}{94.9\angle 18.4^\circ}$$
$$= 61.46\angle -24.3^\circ = 56 - j25.3$$

The circuit now looks like this,

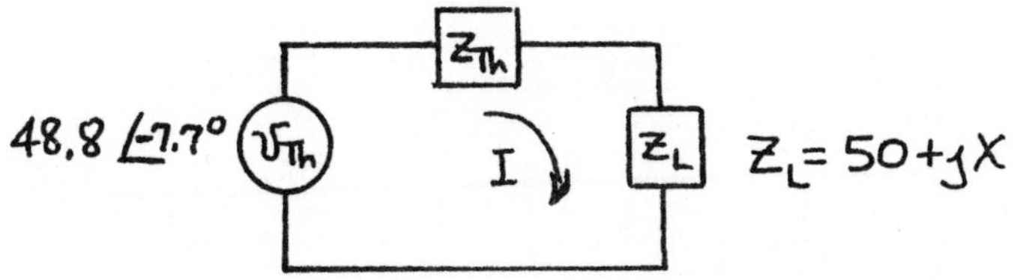

The phase angle of the source is only a reference to the phase of the original source and may be ignored while finding *jX*.

To maximize power, $Z_L = Z^*_{Th}$, that is, the load impedance should be the complex conjugate of the source. Since we cannot make the real parts equal, $56 \neq 50$, we must recognize that the maximum current through R_L will deliver maximum power.

$$I = \frac{V_{Th}}{Z_{Th} + Z_L}$$

$$Z_{Th} + Z_L = 56 - j25.3 + 50 + jX$$

The denominator will be minimum when $|-j25.3| = |+jX|$

then $X = \omega L = 25.3$

$$L = \frac{25.3}{\omega} \quad \text{but } \omega = 2\pi(60) = 377$$

$$L = \frac{25.3}{377} = 67.2 \text{ millihenrys}$$

$$|I| = \frac{48.8}{50+56} = \frac{48.8}{106} = 0.46 \text{ Amps Rms}$$

So $\text{Power} = I^2 R_L$

$$P = (0.46)^2(50) = 10.6 \text{ watts}$$

CIRCUITS 6

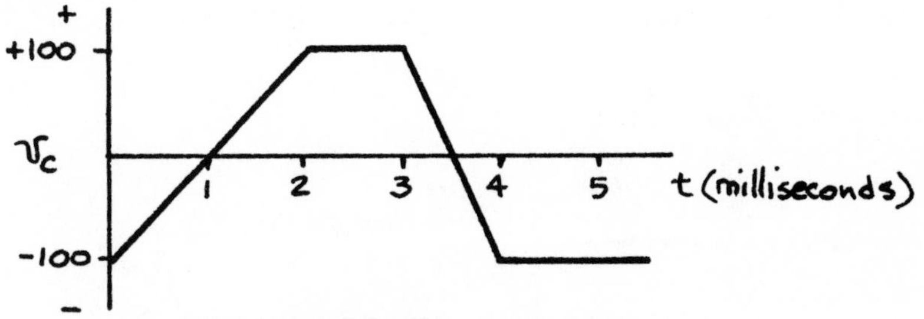

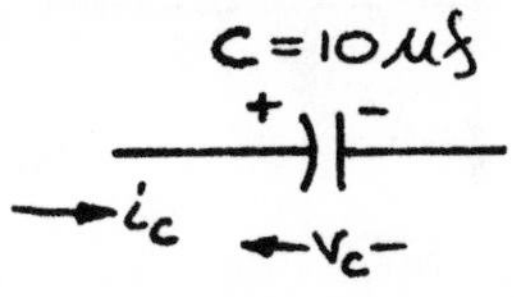

The voltage waveform shown occurs across the 10μf capacitor. Sketch and dimension the proper waveform for the current through the capacitor.

Could this waveform occur if the capacitor were in series with a 1 henry inductor? Explain.

SOLUTION

The basic relationship between voltage and current in a capacitor is, $i_c = C\frac{dv}{dt}$.

From the waveform between 0 and 2 milliseconds, V rises linearly from -100V to +100V, or $\Delta V = 200V$. So $\frac{dv}{dt} = \frac{200}{2\times10^{-3}}$

$$\frac{dv}{dt} = 10^5 \text{ Volts/sec} \quad 0<t<2\text{ ms}$$

$$i = C\frac{dv}{dt} = 10\times10^{-6}\times10^5 = 1 \text{ amp}$$

then for $2<t<3$ ms, V is constant at +100V

$$\text{So } \frac{dv}{dt} = 0 \quad \text{and } i = 0 \text{ amps}$$

For $3<t<4$ msec, $\Delta V = -100-(100) = -200V$

$$\frac{dv}{dt} = \frac{-200}{10^{-3}} = -2\times10^5$$

$$\text{and } i = 10\times10^{-6}(2\times10^5) = -2 \text{ amps}$$

For $t > 4$ msec, $V =$ constant $= -100$ Volts

$\frac{dv}{dt} = 0, \quad i = 0$ amps

Sketching:

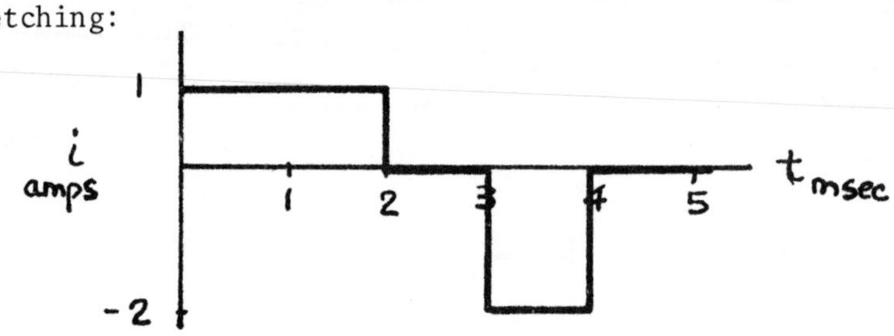

The current consists of a 1 amp pulse between 0 and 2 ms and a -2 amp pulse between 3 and 4 ms, and zero elsewhere.

Since the voltage across an inductor is $L\frac{di}{dt}$ and these pulses represent $\frac{di}{dt} \rightarrow \infty$ at points of change, only impulses of voltage (very large values) could approximate this current through an inductor. Thus you could say that such a current waveform could not occur through an inductor with finite voltage sources.

CIRCUITS 7

Consider the following expression for a wave:

$f(t) = 10.0 \sin wt + 2.0 \cos (3wt + 90°)$

(1) Wt. 3 On the grid provided sketch the wave, f(t) to the scale given by locating ordinates for every 30° on the wt scale.

(Example) For the statement: The maximum value of the fundamental component of the wave, f(t), is:

a. - 0.0 c. - 10.0
b. - 2.0 d. - None of these

The designation, "c." for 10.0, should be checked.

(2) Wt. 2 The d-c component of the wave, f(t), is:

a. - 0.0 f. - 3.0
b. - 1.0 g. - 4.0
c. - 1.5 h. - 5.0
d. - 2.0 i. - 10.0
e. - 2.5 j. - None of these

(3) Wt. 3 The half-period average of the wave, f(t), is:

a. - 0.0 f. - 6.36
b. - 1.20 g. - 7.64
c. - 2.00 h. - 10.0
d. - 3.28 i. - 12.0
e. - 5.96 j. - None of these

(4) Wt. 2 The rms or effective value of the wave, f(t), is:

a. - 2.0 f. - 7.63
b. - 3.18 g. - 8.02
c. - 6.36 h. - 9.96
d. - 7.07 i. - 11.3
e. - 7.21 j. - None of these

SOLUTION

(1) $f(t) = 10.0 \sin\omega t + 2.0 \cos(3\omega t + 90°)$
To determine values of the second term, set up a table:

ωt	$3\omega t$	$(3\omega t+90°)$	$\cos(3\omega t+90°)$
0	0	$\pi/2$	0
$\pi/6$	$\pi/2$	π	-1
$\pi/3$	π	$3\pi/2$	0
$\pi/2$	$3\pi/2$	2π	+1

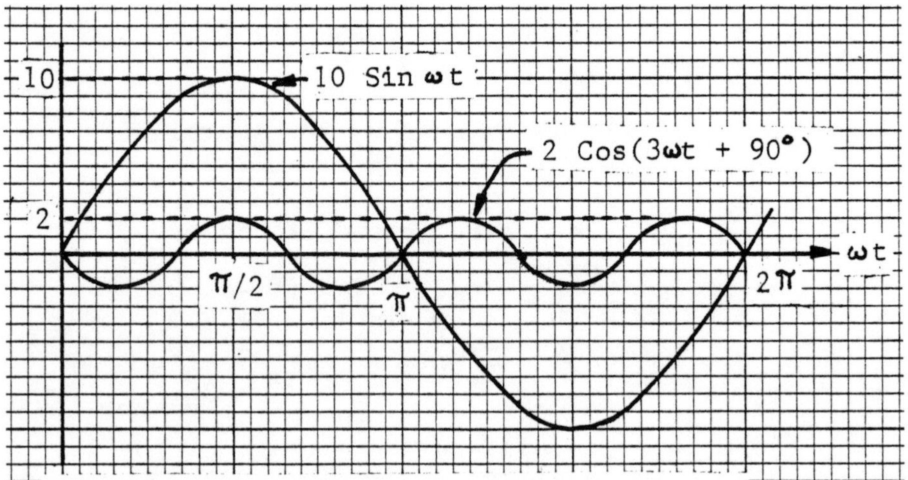

Then f(t) is the sum of these two waves:

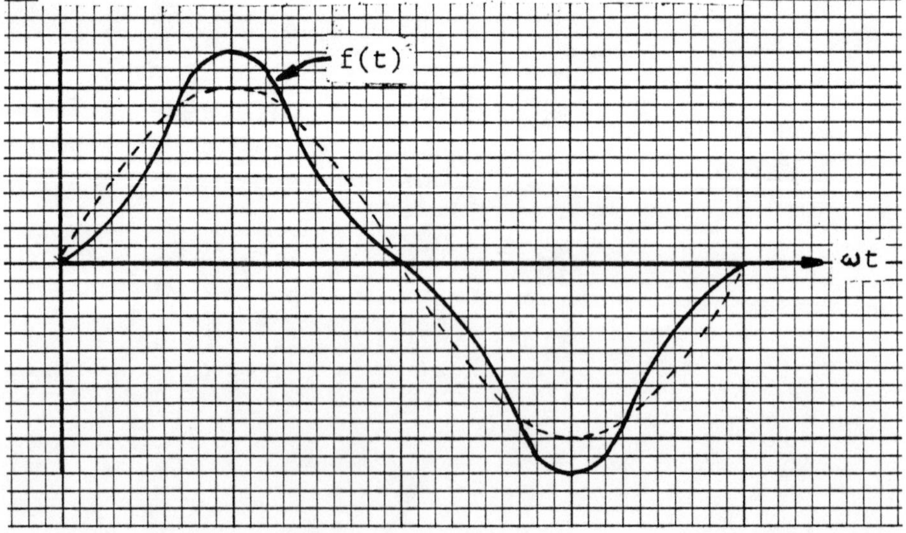

(2) The d.c. component:

$$\text{d.c. component} = a_0 = \frac{1}{2\pi}\int_0^{2\pi} f(t)\ d(\omega t)$$

$$= \frac{1}{2\pi}\int_0^{2\pi} 10\text{Sin}\ \omega t\ d(\omega t) + \frac{1}{2\pi}\int_0^{2\pi} 2\text{Cos}(3\omega t+90°)\ d(\omega t)$$

= 0+0 = 0. The answer is a. - 0.0

(3) The half-period average of the wave, f(t), is:

$$\text{d.c. half-period average} = a_{T/2} = \frac{1}{\pi}\int_0^{\pi} f(t)\ d(\omega t)$$

$$= \frac{1}{\pi}\left[\int_0^{\pi} 10\text{Sin}\ \omega t\ d(\omega t) + \frac{1}{3}\int_0^{\pi} 2\text{Cos}(3\omega t+90°)\ d(3\omega t)\right]$$

$$= \frac{1}{\pi}\left[+10 + 10 - 2/3 - 2/3\right] = \frac{1}{\pi}\left[20 - 4/3\right] = \frac{18.67}{\pi}$$

= 5.95 The answer is e. - 5.96

Note that there is one more negative than positive third harmonic loop per half-cycle.

(4) The rms or effective value of the wave, f(t), is:

$$\text{effective value of fundamental} = F_1 = \frac{10}{\sqrt{2}}$$

$$\text{effective value of 3rd harmonic} = F_3 = \frac{2}{\sqrt{2}}$$

$$F_{TOT(effective)} = \sqrt{F_1^2 + F_3^2} = \sqrt{\frac{100}{2} + \frac{4}{2}} = 7.21$$

The answer is e. - 7.21

CIRCUITS 8

Consider again the expression for the wave in Problem 7 as follows:

f(t) = 10.0 sin wt + 2.0 cos (3wt + 90°)

(1) Wt. 2 The maximum or peak value of the wave, f(t), is:

a. - 2.0
b. - 4.0
c. - 6.0
d. - 8.0
e. - 10.0
f. - 11.5
g. - 12.0
h. - 14.0
i. - 16.0
j. - None of these

(2) Wt. 2 If the wave, f(t), represents a current in amperes that is flowing through a resistor having a resistance of 3.00 ohms, the power loss in watts is:

a. - 43
b. - 55
c. - 78
d. - 93
e. - 110
f. - 156
g. - 187
h. - 221
i. - 312
j. - None of these

(3) Wt. 3 If the wave, f(t), represents a current in amperes that is flowing through an inductor having an inductance of 0.1 henry, the rms or effective voltage in volts across the inductor is:

a. - 0.00
b. - 0.33
c. - 0.67
d. - 0.83
e. - 1.00
f. - 0.33w
g. - 0.67w
h. - 0.83w
i. - 1.00w
j. - None of these

(4) Wt. 3 If the wave, f(t), represents a current in amperes having an angular velocity of 6000 radians per second which current is passing through a capacitor having negligible losses, no initial charge, and a capacitance of 0.002 farads; the rms or effective voltage in volts across the capacitor is:

a. - 0.0
b. - 0.21
c. - 0.59
d. - 0.74
e. - 1.00
f. - 2.10
g. - 5.90
h. - 7.40
i. - 10.0
j. - None of these

SOLUTION

(1) Note from the sketch the fundamental and the 3rd harmonic peaks occur in phase, thus $f(t)_{peak}$ is merely:

$$f(t)_{peak} = 10 + 2 = 12$$

The answer is g. - 12.0

(2) From Problem 7, part (4):

$$I_{eff.} = \sqrt{I_1^2 + I_3^2} = 7.21 \text{ amperes}$$

then:

$$P = I_{eff}^2 R = (7.21)^2(3.00) = 156 \text{ watts}$$

The answer is f. - 156

(3) The effective voltage across the inductor:

$$E_{eff.} = \sqrt{E_1^2 + E_3^2}$$

where $E_1 = I_1 X_{L1} = \left(\frac{10}{\sqrt{2}}\right)(\omega L) = \frac{1.0}{\sqrt{2}}\omega$

and $E_3 = I_3 X_{L3} = \left(\frac{2}{\sqrt{2}}\right)(3\omega L) = \frac{0.6}{\sqrt{2}}\omega$

$$\therefore E_{eff.} = \sqrt{\frac{\omega^2}{2} + \frac{.36\omega^2}{2}} = \frac{\omega}{\sqrt{2}}\sqrt{1 + .36} = 0.83\omega$$

The answer is h. - 0.83 ω

(4) For ω = 6000 radians/second:

$$E_1 = X_{C1} I_1 = \left(\frac{1}{\omega C}\right)\left(\frac{10}{\sqrt{2}}\right) = \frac{10}{(6\text{x}10^3)(2\text{x}10^{-3})\sqrt{2}} = \frac{10}{12\sqrt{2}}$$

$$E_3 = X_{C3} I_3 = \left(\frac{1}{3\omega C}\right)\left(\frac{2}{\sqrt{2}}\right) = \frac{2}{3\text{x}6\text{x}10^3\text{x}2\text{x}10^{-3}} = \frac{2}{36\sqrt{2}}$$

$$\therefore E_{eff.} = \sqrt{\left(\frac{10}{12\sqrt{2}}\right)^2 + \left(\frac{2}{36\sqrt{2}}\right)^2} = 0.59$$

The answer is c. - 0.59

CIRCUITS 9

CIRCUITS: TRANSIENTS

A 2.0 μf capacitor is charged so that it has 100 volts across its terminals. The terminals are suddenly connected through a negligible resistance (actually by means of copper bars, 1/2" square and 10" long) to the terminals of a 4.0 μf capacitor having no initial charge.

(1) Wt. 2 What are the final steady state voltages across each capacitor?

(2) Wt. 1 What are the initial stored energies of each capacitor?

(3) Wt. 1 What are the final steady state stored energies of each capacitor?

A 100 ohm resistor is placed in series with the 4.0 μf capacitor so that the charging current will flow through the resistor. The same 2.0 μf capacitor charged to 100 volts is connected to the combination of the 4.0 μf capacitor and 100 ohm resistor in series with the 4.0 μf capacitor having no initial charge.

(4) Wt. 2 What is the time constant of the circuit?

(5) Wt. 1 What are the final steady state voltages across each capacitor?

(6) Wt. 1 What are the final steady state stored energies of each capacitor?

(7) Wt. 2 How do you explain or account for the results of the energies calculated in (3) and (6) above in the light of the different resistor power losses?

SOLUTION

$Q = CE = 2 \times 10^{-6} \times 100 = 200 \times 10^{-6}$ coulomb

Capacitors are suddenly connected together, assuming no circuit resistance and the 4 μfd capacitor has $Q_2 = 0$.

The capacitors are in parallel across V so electrons will flow until equilibrium is reached.

The total charge $Q = 200 \times 10^{-6}$ coulomb remains in the system.

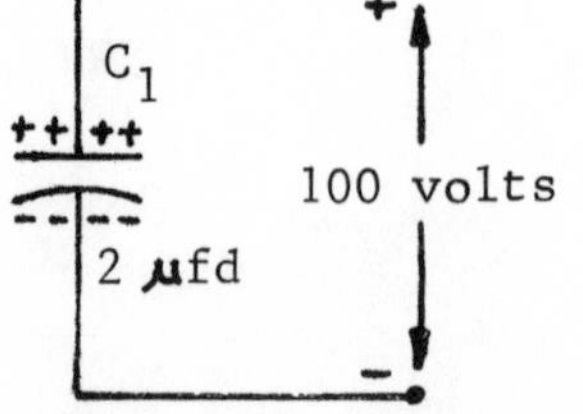

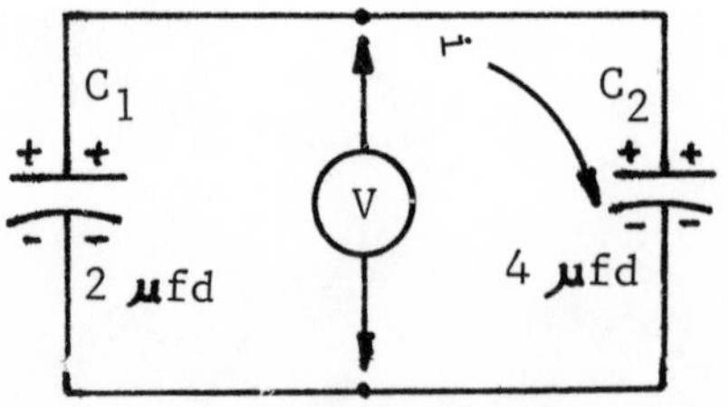

So $V = \frac{Q}{C_1 + C_2} = \frac{200 \times 10^{-6}}{(2 + 4)\times 10^{-6}} = 33\ 1/3$ Volts

(1) The voltage across each capacitor is 33 1/3 volts.

(2)

$$W = \int \frac{1}{C}\, dq = \frac{1}{2}\frac{Q^2}{C} \text{ and initially } 0$$

$Q_1 = 200 \times 10^{-6}$ coulomb so:

$$W_1 = \frac{1}{2}\frac{(200 \times 10^{-6})^2}{2 \times 10^{-6}} = \frac{(2 \times 10^{-4})^2}{4 \times 10^{-6}}$$

$$W_1 = \frac{1 \times 10^{-8}}{10^{-6}} = 1 \times 10^{-2} \text{ Joules}$$

energy initially in C_1

Q_2 was zero initially, so $W_2 = 0$

(3) Final energy

$Q_1 = 2 \times 10^{-6} \times 33\ 1/3 = 66\ 2/3 \times 10^{-6}$ coulomb

$$W_1 = \frac{1}{2}\frac{(66\ 2/3 \times 10^{-6})^2}{2 \times 10^{-6}} = \frac{4.45 \times 10^{-10}}{4 \times 10^{-6}}$$

$= 1.110 \times 10^{-3}$ Joules in C_1

$Q_2 = 4 \times 10^{-6} \times 33\ 1/3 = 133\ 1/3 \times 10^{-6}$

$$W_2 = \frac{1}{2}\frac{(133\ 1/3 \times 10^{-6})^2}{4 \times 10^{-6}} = \frac{1.76 \times 10^{-8}}{8 \times 10^{-6}}$$

$= 2.22 \times 10^{-3}$ Joules in C_2

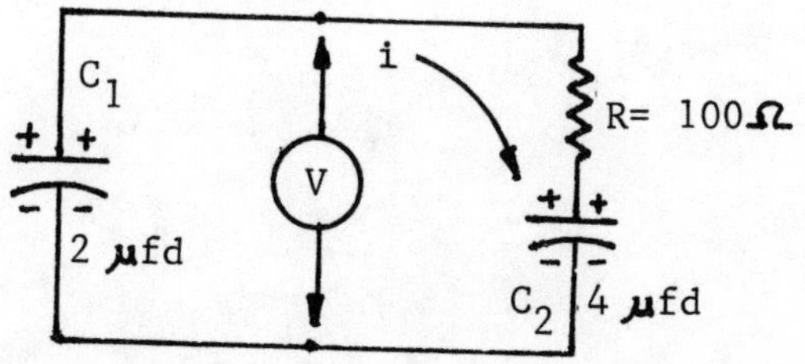

Time Constant t = RC

$$Ri + \frac{1}{C_1}\int i\,dt + \frac{1}{C_2}\int i\,dt = 0$$

i steady state = 0, so:
i transient is the solution of

$$R\frac{di}{dt} + \frac{1}{C_1}i + \frac{1}{C_2}i = 0 \qquad R\frac{di}{dt} + \left(\frac{C_1 + C_2}{C_1C_2}\right)i = 0$$

$$\text{so } i = K\epsilon^{-\left(\frac{C_1 + C_2}{C_1C_2R}\right)t} \qquad \text{where } \frac{C_1 + C_2}{C_1C_2R} = \frac{2 + 4}{2\text{x}4R} = \frac{3}{4R}$$

$$i = \frac{V_{C_1}}{R}\epsilon^{-\left(\frac{C_1 + C_2}{C_1C_2R}\right)t} \qquad \text{so } \frac{t}{RC} = \frac{t}{100\text{x}\frac{4}{3}} = 1$$

(4) Time Constant t = RC = $100 \times 1.33 \times 10^{-6}$
= 133×10^{-6} Seconds

(5) $E_{C_1} = E_{C_2}$ = 33 1/3 Volts, same as part (1).

(6) $$W_1 = \frac{1}{2}\frac{(66\ 2/3 \times 10^{-6})^2}{2 \times 10^{-6}} = 1.11 \times 10^{-3} \text{ Joules}$$

$$W_2 = \frac{1}{2}\frac{(133\ 1/3 \times 10^{-6})^2}{4 \times 10^{-6}} = 2.22 \times 10^{-3} \text{ Joules}$$

(7) The circuit resistance determines the peak discharge or charge current, so there is I^2R loss in parts 1, 2, & 3 even if the resistance of the copper bar seems negligible. This accounts for the loss in energy.

CIRCUITS 10

The circuit shown below is in a steady state.

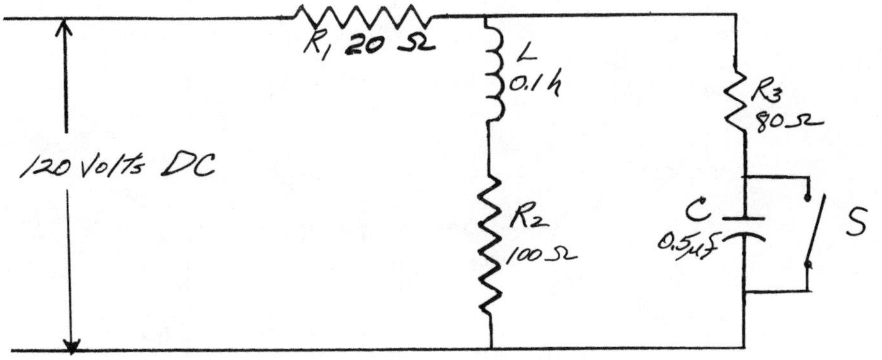

REQUIRED:

Wt.		
1	(a)	What current is drawn from the power source in this initial steady state condition?
9	(b)	What is the analytical expression for the current drawn from the power source after closing switch S?

SOLUTION

a) With the switch open, and with the statement that the circuit is in steady state (to a dc source), one may make the assumption that the current through the inductor is no longer changing; the voltage across this element will then be zero. The voltage across the capacitor has built up to its steady state value and therefore no current will be flowing in this branch:

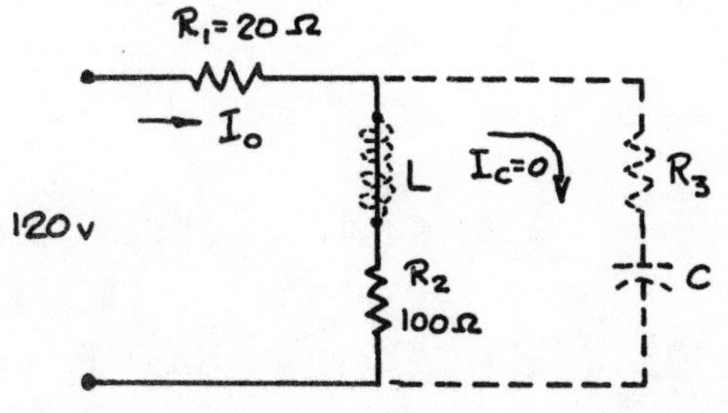

$$I_o = \frac{E}{R_1 + R_2} = \frac{120}{120} = 1 \text{ A.}$$

b) After the switch is closed, the equivalent circuit will be as follows (with an initial inductor current as found in part a):

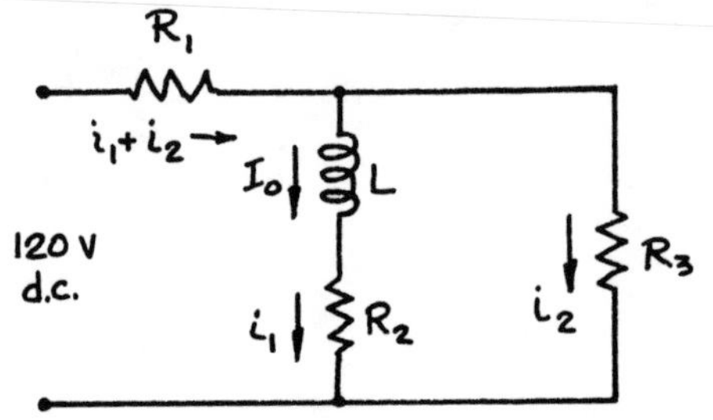

Loop (differential) equations:

1) $E = R_1(i_1 + i_2) + L\dfrac{di_1}{dt} + R_2 i_1$

2) $E = R_1(i_1 + i_2) + R_3 i_2$

with an initial condition of $i_1(0) = I_0$

Using Laplace transforms to solve:

1) $\dfrac{E}{S} = R_1(I_1 + I_2) + L\left[sI_1 - i_1(0)\right] + R_2 I_1$

2) $\dfrac{E}{S} = R_1(I_1 + I_2) + R_3 I_2$

Rearranging terms:

1) $\dfrac{E}{S} + L\,i_1(0) = \left[(R_1 + R_2) + Ls\right] I_1 + R_1 I_2$

$\dfrac{120}{S} + 0.1 = (120 + 0.1S)\, I_1 + 20 I_2$

2) $\dfrac{120}{S} = (20)\, I_1 + (100)\, I_2$

Solving for I_1 and I_2:

$$I_1 = \frac{\begin{vmatrix} \left(\frac{120}{S} + 0.1\right) & 20 \\ \left(\frac{120}{S}\right) & 100 \end{vmatrix}}{\begin{vmatrix} (120 + 0.1S) & 20 \\ 20 & 100 \end{vmatrix}} = \frac{\left(\frac{120}{S} + 0.1\right)100 - \left(\frac{120}{S}\right)20}{(120 + 0.1S)\,100 - 20^2}$$

$$= \frac{96 + 0.1S}{S(116 + 0.1S)}$$

$$I_2 = \frac{\begin{vmatrix} (120 + 0.1S) & \left(\frac{120}{S} + 0.1\right) \\ 20 & \left(\frac{120}{S}\right) \end{vmatrix}}{(120 + 0.1S)100 - 20^2} = \frac{120 + 0.1S}{S(116 + 0.1S)}$$

But $I_{source} = I_1 + I_2$

$$\therefore\ I_s = \frac{96 + 0.1S + 120 + 0.1S}{S(116 + 0.1S)} = \frac{216 + 0.2S}{S(116 + 0.1S)}$$

$$= \frac{216}{116}\left[\frac{1 + \frac{0.2}{216}S}{S\left(1 + \frac{0.1}{116}S\right)}\right] = 1.86\left[\frac{1 + \tau_1 S}{S(1 + \tau_2 S)}\right]$$

where $\tau_1 = 0.000927$

$\tau_2 = 0.0008625$

$$\therefore\ i_s(t) = \mathcal{L}^{-1}[I_s] = 1.86\left[1 - (1 - \tau_1/\tau_2)e^{-t/\tau_2}\right]$$

$$= 1.86\,(1 + 0.072\,e^{-t/\tau_2})$$

CIRCUITS 11

Refer to the circuit below. If the inductance L is 10.0 microhenries and the frequency of the applied voltage is 9.55 megacycles, determine the value of the reactance of C so that the circuit will be series resonant. What is the impedance looking into the circuit under these conditions?

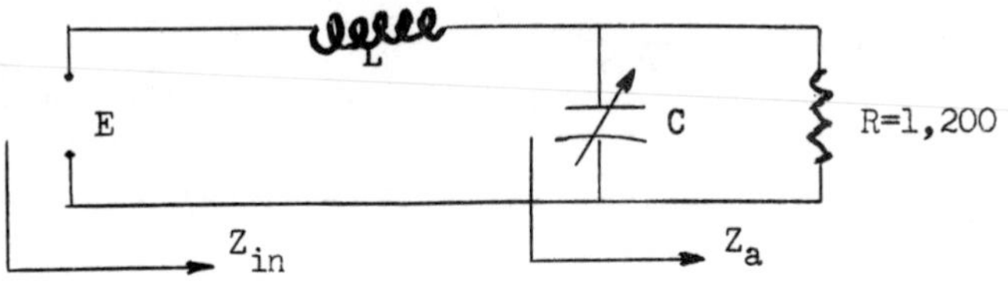

SOLUTION

(a) Solve circuit by the admittance method:

$$Y_a = \frac{1}{Z_a} = \frac{1}{1{,}200} + jYca = 0.833 \times 10^{-3} + jYca$$

$$Z_a = \frac{1}{0.833 \times 10^{-3} + jYca} = \frac{0.833 \times 10^{-3}}{D} - \frac{jYca}{D} = R_a - jXca$$

Where denominator = D after rationalizing the fraction with the conjugate ; $D = 0.693 \times 10^{-6} + Y_{ca}^2$,

using: $(a-b)(a+b) = a^2 - b^2$.

New equivalent circuit:

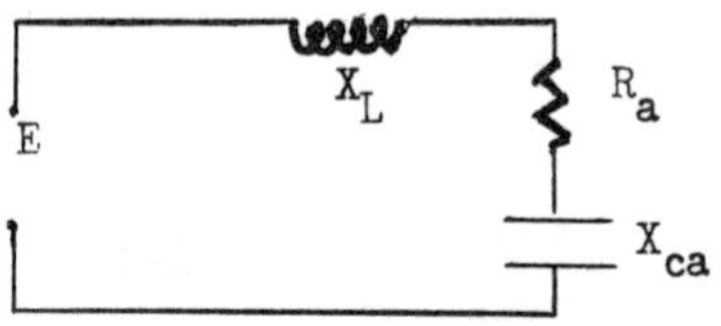

At Resonance: $X_{ca} = X_L$

$$X_L = 2\pi fL = (6.28)(9.55 \times 10^6)(10 \times 10^{-6}) = 599.7 \text{ ohms}$$

Thus $\frac{Yca}{D} = 599.7 = X_{ca}$ based on the principles of resonance.

If $\frac{Yca}{D} = 599.7$, then

$$Y_{ca} = 599.7\,(0.693 \times 10^{-6} + Y_{ca}^2)$$

$$599.7Y_{ca}^2 - Y_{ca} + 4.156 \times 10^{-4} = 0$$

Solving the second degree equation by using the standard formula we obtain:

$$Y_{ca} = \frac{1 \pm \sqrt{1 - 4\ (4.156 \times 10^{-4})\ (599.7)}}{2 \times 599.7}$$

$$= \frac{1 \pm \sqrt{1 - 0.9979}}{2 \times 599.7} = \frac{1}{1,199.4} = 0.833 \times 10^{-3}$$

Where the radical was approximated to be zero

Therefore the reactance requested is:

$$X_{ca} = \frac{1}{Y_{ca}} = 1,199.4 \text{ ohms}$$ ANS.

(b) $Z_{in} = R_a$, as in resonance the only effective part of the impedance is the real part of the complex expression

$$Z_{in} = R_a = \frac{0.833 \times 10^{-3}}{D} = \frac{0.833 \times 10^{-3}}{0.693 \times 10^{-6} + Y_{ca}^2} =$$

$$\frac{0.833 \times 10^{-3}}{0.693 \times 10^{-6} + (0.833 \times 10^{-3})^2}$$

$$Z_{in} = 0.6 \times 10^3 = 600 \text{ ohms}$$ ANS.

CIRCUITS 12

The switch was closed sufficiently long ago that the current i (through the source) has reached steady state. The switch is then opened at time $t = t_o$.

REQUIRED:

Wt

1 (a) Find the current i just before switch is opened; that is, at $t = t_o-$

3 (b) Find the current i just after the switch is opened; that is, at $t = t_o+$

6 (c) Find the current i as a function of time after the switch is opened; that is, find $i\ (t - t_o)$.

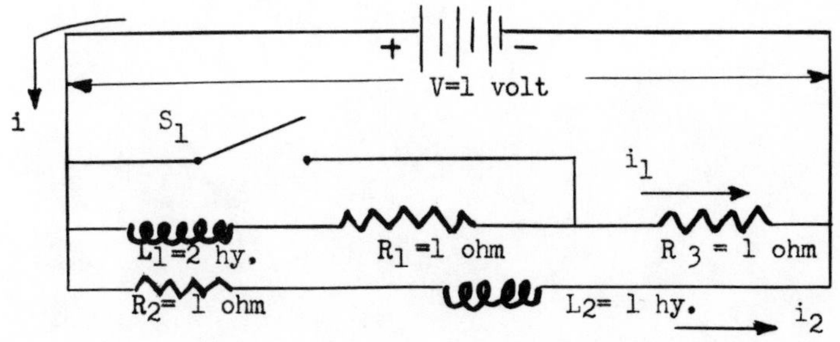

SOLUTION

(a) At $t = t_o-$ only R_3 and R_2 determine the magnitude of i

$$R_{total} = \frac{1 \times 1}{1 + 1} = 0.5 \text{ ohm}$$

$$i = \frac{E}{R_{total}} = \frac{1}{0.5} = 2 \text{ amp}$$ ANS.

(b) At $t = t_o+$, since the current in an inductance will not change instantaneously, the current in L_1 will remain at 0. Only R_2 will determine the magnitude of i

$$i = \frac{E}{R_2} = \frac{1}{1} = 1 \text{ amp}$$ ANS.

(c) At $(t - t_o)$ the R_2 and L_2 branch of the circuit is in steady state and is not considered. Therefore for the L_1, R_1 and R_3 branch:

$$i = i_{transient} + i_{R_2}$$

$$i_{transient} = \frac{E}{R} \left(1 - e^{-\frac{Rt}{L_1}} \right)$$

i_{R_2} = 1 amp (See above (b))

$$R = R_1 + R_3 = 1 + 1 = 2 \text{ ohms}$$

Therefore

$$i = \frac{1}{2} - \frac{1}{2} e^{-t} + 1 = 1.5 - 0.5e^{-t}$$ ANS.

Proof: at t = o .... $e^{-t} = 1.0$ and $i = 1.5 - 0.5 \times 1.0 =$ 1 amp, q.e.d.

NOTE: Part (c) can be also worked by using the classic solutions of:

$$L_1 \frac{di_1}{dt} + (R_1 + R_3)\, i_1 = 1$$

$$L_2 \frac{di_2}{dt} + R_2\, i_2 = 1$$

$i = i_1 + i_2 = 1.5 - 0.5e^{-t}$ by applying the proper boundary conditions.

CIRCUITS 13

An alternating current voltmeter consists of a series connection of an ideal half-wave diode and a D'Arsonval meter. The meter is calibrated to read the rms value of an applied voltage. When the waveform sketched below is applied, the meter reads 80 volts. What is the peak value of the applied waveform?

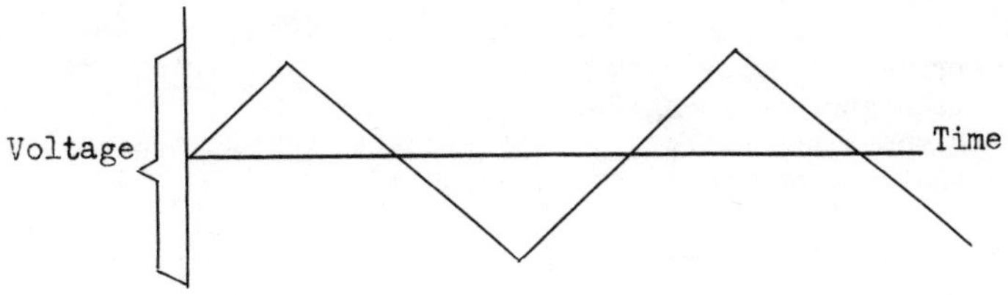

SOLUTION

The voltage as seen by the D'Arsonval meter (assuming a lossless diode rectifier) from a pure sine wave: $e = E_{max} \sin \omega t$

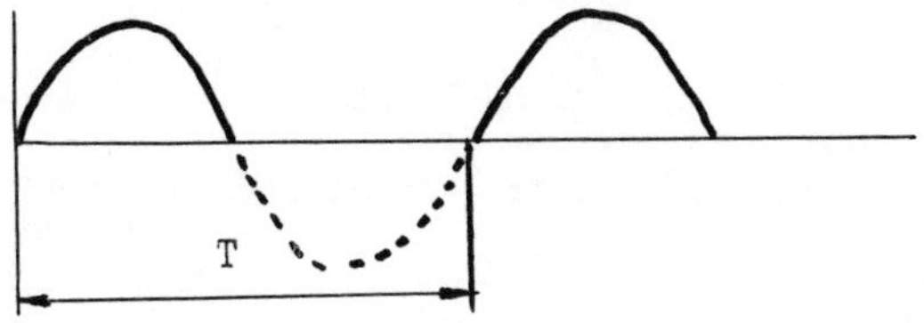

Then: $E_{avg} = \frac{1}{T}\int_{0}^{\frac{1}{2}T} E_{max} \sin\omega t\, d(\omega t) + \frac{1}{T}\int_{\frac{1}{2}T}^{T} o\, d\omega t$

Actual voltage read would be E_{max}/π but meter is calibrated to read $E_{max}/\sqrt{2}$ (for an rms value).

For the wave shape given:

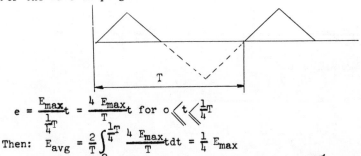

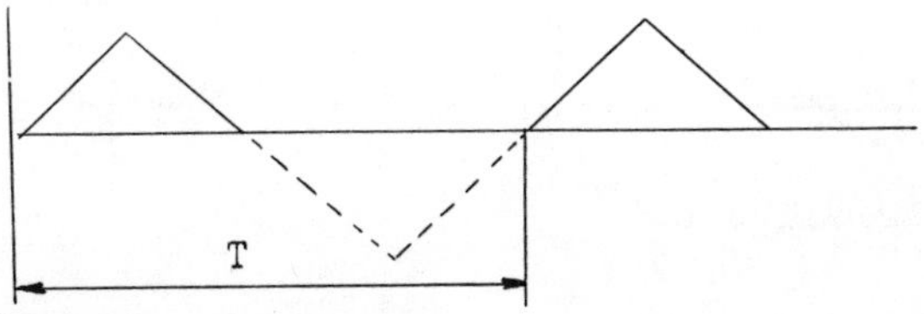

$e = \frac{E_{max}}{\frac{1}{4}T} t = \frac{4\,E_{max}}{T} t$ for $o < t < \frac{1}{4}T$

Then: $E_{avg} = \frac{2}{T}\int_{0}^{\frac{1}{4}T} \frac{4\,E_{max}}{T} t\,dt = \frac{1}{4} E_{max}$

But meter reads (for a pure sine wave): $E_{meter} = \frac{\pi}{\sqrt{2}} E_{avg}$

Then, for the saw-tooth wave, as given: 80^{x} volts $= \frac{\pi}{\sqrt{2}} E_{avg} = \frac{\pi}{\sqrt{2}} (\frac{1}{4} E_{max})$

$E_{max} = (\frac{\sqrt{2}}{\pi})\,(4)\,(80) = 143.7$ volts ANS.

^{x}NOTE: 80 volts is not the true rms voltage of the given wave form.

CIRCUITS 14

REQUIRED:

Wt.		
5	(a)	In the bridge circuit shown below, find the current through the meter as a function of R, the deviation of the unknown resistor from 1000 ohms.
3	(b)	Find the points at which the meter should be marked if the bridge is to be used to reject resistors which exceed 10% tolerance ($\lvert R \rvert \leq 100$).
2	(c)	Discuss the effect of battery aging on the tolerance, assuming that the marks are not readjusted.

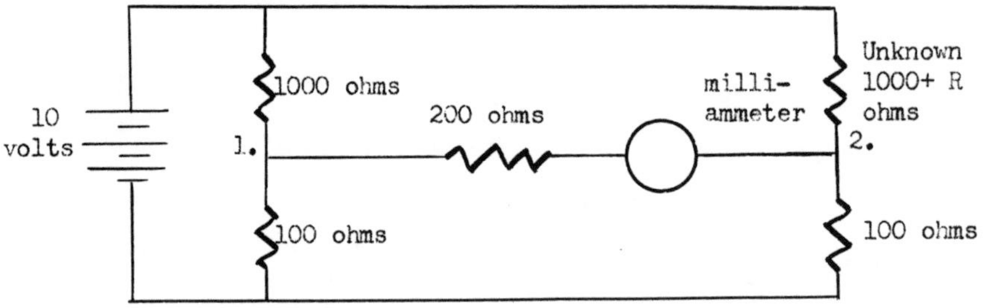

SOLUTION

(a) Based on the principle of voltage dividers (for finding Thevenin's open circuit voltage):

$$V_1 = \frac{100 \times 10}{1{,}100} \text{ and } V_2 = \frac{100 \times 10}{100 + P}$$

Where P = 1000 + R

$$V_{12} = V_1 - V_2 = 10^3 \left(\frac{1}{1{,}100} - \frac{1}{100 + P}\right) = 10^3 \left[\frac{100 + P - 1{,}100}{1{,}100\ (100 + P)}\right]$$

$$= 10^3 \left[\frac{P - 1000}{1{,}100\ (P + 100)}\right] = \frac{P - 1000}{1.1\ (P + 100)}$$

Based on Thevenin theorem:

$$Z_{12} = \frac{100 \times 1{,}000}{1{,}100} + \frac{100\ P}{100 + P} = \frac{10 \times 10^6 + 10^5 P + 11 \times 10^4 P}{1{,}100\ (100 + P)}$$

$$= \frac{10^3\ (10 + 0.21P)}{110 + 1.1P}$$

$$Z_{12} + 200 = \frac{10{,}000 + 210P + 22{,}000 + 220P}{110 + 1.1P}$$

$$= \frac{430P + 32{,}000}{1.1P + 110} = \frac{430P + 3{,}200}{1.1\ (P + 100)}$$

$$I_a = \frac{V_{12}}{Z_{12} + 200} = \frac{P - 1000}{1.1\ (P + 100)} \times \frac{1.1\ (P + 100)}{430P + 32{,}000} = \frac{P - 1000}{430P + 32{,}000}$$

Substituting 1,000 + R = P we obtain:

$$I_a = \frac{R}{430\ R + 462{,}000}$$ ANS.

(b) If R= +100

Mark one meter point: $I_a = \frac{100}{43{,}000 + 462{,}000} = \frac{100}{505{,}000}$

$= 0.198 \times 10^{-3}$ Amps ANS.

If R = -100

Mark second meter point: $I_a = \frac{-100}{-43{,}000R + 462{,}000} = \frac{-100}{41{,}900}$

$= -0.238 \times 10^{-3}$ Amps ANS.

(c) As the battery ages, its internal resistance increases, reducing all voltages including the voltage differences between points 1 and 2 across which the current is being measured. The range of acceptable current readings would be therefore smaller. If the marks are not reduced accordingly, some resistors would be accepted that should be rejected. ANS.

3

Electro-Magnetic Fields

Problems in this section are generally of three types: transmission lines including both co-axial and parallel wire forms, magnetic circuits, and wave propagation in various media.

Transmission line problems are solved most easily with the aid of the Smith chart and a few basic relationships. Important parameters for transmission lines are: the characteristic impedance, Z_0; the complex propagation constant, γ (Gamma); the voltage reflection coefficient, ρ_v; and the current reflection coefficient, ρ_i.

These parameters may or may not be given explicitly, however they are usually easily calculated from available data.

The distributed series impedance of a line, Z, in ohms per meter, is a complex quantity $R + j\omega L$ while the distributed shunt admittance, Y, is $G + j\omega C$ mhos per meter. The characteristic impedance of the line then is $Z_0 = \sqrt{Z/Y}$. When $R << |j\omega L|$ and $G << |j\omega C|$, the line is assumed to be lossless and Z_0 becomes $\sqrt{L/C}$, a pure resistance. In this expression, L is the inductance per unit length and C is the capacitance per unit length. Lossless lines can easily be manipulated on a Smith chart.

Gamma (γ), the complex propagation constant is given by $\gamma = \sqrt{ZY}$, or $\gamma = \alpha + j\beta$. α (alpha) is the real part of the constant and is called the attenuation constant with units of nepers/meter. β (beta) is the imaginary part of the constant and is called the phase constant, radians/meter.

When transmission lines are terminated with an impedance, Z_L, where Z_L is complex and may be anything from 0 (short circuit) to infinity (open circuit), an important property of the line called the reflection coefficient is created.

Classical transmission line theory has energy propagating in two waves, the incident wave which propagates from the source toward the load, and the reflected wave which propagates from the load back to the source.

The reflection coefficient for voltage is given by $\rho_v = \dfrac{Z_L - Z_o}{Z_L + Z_o}$ and for current by $\rho_i = \dfrac{Z_o - Z_L}{Z_L + Z_o}$.

Note that $\rho_v = -\rho_i$. Further note that for $Z_L = 0$ (short circuit), $\rho_v = -1$ and for $Z_L = \infty$ (open circuit), $\rho_v = +1$. A very important measure of the transmission line's effectiveness is called the Voltage Standing Wave Ratio or *VSWR*. The *VSWR* is given by:

$$VSWR = \frac{1 + |\rho_v|}{1 - |\rho_v|}$$

Again, note here that either a short circuit or open circuit termination produces an infinite *VSWR*. In addition, a line terminated in its characteristic impedance, $Z_L = Z_o$, has a zero reflection coefficient and a *VSWR* of unity.

An excellent reference for transmission line problems including a good illustration of Smith chart use is, *Electromagnetics* by John D. Kraus and Keith R. Carver, 2nd ed., 1973. McGraw-Hill.

Magnetic circuit problems are of two basic types; one, calculating the flux density in a magnetic circuit, or two, calculating the force on a magnetic pole.

Magnetic flux problems can be treated by the magnetic circuit analogy to Ohm's law. In this analogy, the magnetizing force, Ampere Turns (*NI*) is analagous to voltage in Ohm's law. Magnetic reluctance $(\ell/A\mu)$ is analagous to resistance. (ℓ is the mean length of the path, *A* is the cross sectional area of the path, and μ is the permeability of the material of the path.) Current corresponds to total flux, *BA*, where *B* is the flux density in webers/meter^2 and *A* is the cross-sectional area in meters^2.

Then $NI = (BA)(\Sigma R)$

where ΣR is the sum of the reluctances around the path.

EXAMPLE: An iron ring of mean length ℓ has a small $(g \ll \ell)$ gap (g) cut in it transversely. The cross-sectional area of the ring and gap is A. The required flux density in the gap is B. How many ampere turns are required?

SOLUTION: The reluctance of the ring is $R_R = \frac{(\ell - g)}{A\mu}$, where μ is the total permeability of the iron; and the reluctance of the gap is $R_g = \frac{g}{A\mu_o}$ where μ_o is the permeability of free space, nearly the same for air.

Then the ampere turns needed are,

$$NI = BA\left[\frac{\ell - g}{A\mu} + \frac{g}{A\mu_o}\right]$$

The second major question concerning magnetic circuits is that of the force on a magnetic pole in a magnetic field. An example for the force between the poles in a gap, such as the previous example, will be discussed.

The energy density stored in a magnetic field is

$$E_d = \frac{1}{2}\frac{B^2}{\mu} \text{ Joules/meter}^3$$

If one assumes a gap length of g and a cross sectional area of A, then the total energy stored in the gap is

$$W = \frac{B^2 A g}{2\mu_0}$$

Suppose now, the gap is increased by an infinitesimal amount, $\Delta g \to dg$, then the change in energy stored $\Delta W \to dW$ is given by $dW = \frac{B^2 A}{2\mu_0} dg$. But this change in energy stored represents work done, or force times distance.

Thus $\Delta W = \text{Force} \times \Delta g$

$$\text{So} \quad \text{Force} = \frac{B^2 A}{2\mu_0} \text{ Newtons}$$

References for magnetic circuits and related topics may be found in the Kraus text or *Principles and Applications of Electromagnetic Fields,* by Plonsey and Collin, 1961. McGraw-Hill. Many other good texts on this subject are also available.

On the subject of wave propagation in space or other media and the related topic of antennas, the field is too specialized and would require far more space than is available here to be adequately discussed. Therefore, we recommend that the reader who is qualified in this area consult one of the many texts in this field, including the two previously mentioned sources.

ELECTRO-MAGNETIC FIELDS 1

A transducer whose impedance is 600 ohms resistive generates 0.1 volt rms at a frequency of 1000 Hz. It is desired to transmit this signal over a telephone cable whose length is 2.5 miles. The characteristic impedance of the line, Z_o, is 600 ohms, and its loss is 0.40 neper per mile. The phase shift is 0.1255 radians per mile. All parameters are measured at 1000 Hz. The input to the line is to be amplified by amplifier A so as to feed the line at a level of 0 dbm. The output of the line is to be amplified by amplifier B so as to provide 2.0 watts to a recorder. Amplifiers A and B each have an input impedance of 600 ohms.

REQUIRED:

(a) What is the gain of amplifier A in decibels?

(b) What is the gain of amplifier B in decibels?

(c) What is the delay in microseconds between the input and output of the line?

(d) What would be the least expensive way to reduce the 1000 Hz loss in the telephone line?

(e) What effect, if any, would this measure in (d) above have on the line's delay?

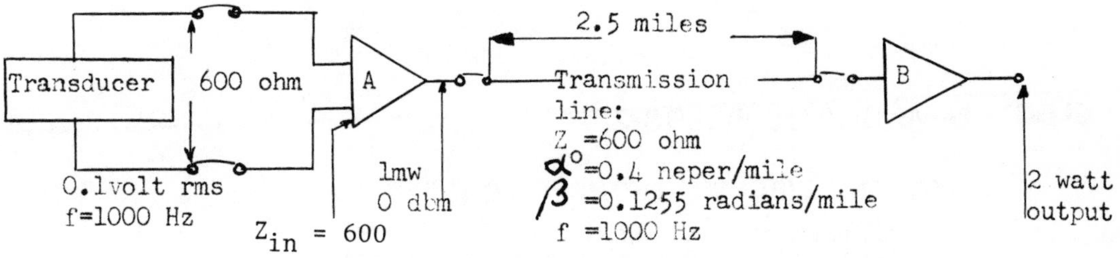

SOLUTION

(a) P_{in} level $= \frac{E^2}{R} = \frac{\left[\frac{1}{2}(0.1)\right]^2}{600} \times 10^3 = \frac{0.0167}{4}$ mw

P_{out} level is given: 0dbm = 1 mw

$$\text{db Gain}_A = 10 \log_{10} \frac{P_{out}}{P_{in}} = 10 \log_{10} \frac{1}{\left(\frac{0.0167}{4}\right)} = 10 \log_{10} 240$$

$$= 10 \times 2.38 = 23.8 \text{ db}$$

(b) Loss in cable: 0.4 nepers/mile x 2.5 miles = 1 neper

1 neper x 8.686 db/nepers = 8.686 db

P_{in} level = 0 - 8.686 = -8.686 dbm

P_{out} level = $10 \log \frac{2}{1 \times 10^{-3}} = 10 \log 2 \times 10^3 = 10 \times 3.3 = 33$ dbm

db $Gain_B$ = 33.0 - (-8.686) = 41.686db ANS.

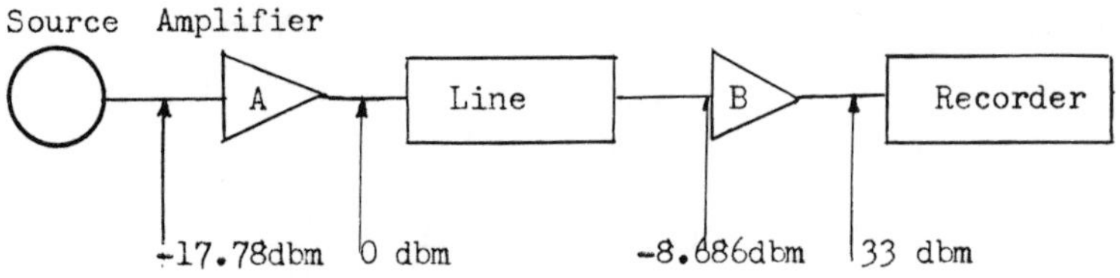

(c) $\text{Delay} = \dfrac{0.1255 \text{ rad/mi} \times 2.5\text{mi} \times 10^6 \frac{\text{microsec}}{\text{sec}}}{2\pi \times 1000 \text{ H}_z} = 50$ microseconds ANS.

(d) The loss at 1000 Hz could be reduced the least expensively by loading (series) the telephone line; i.e., inserting inductance coils at regularly spaced intervals.

(e) The delay of the transmission line will increase when load coils are added, since we increased the phase shift.

ELECTRO-MAGNETIC FIELDS 2

MAGNETIC CIRCUITS: LIFTING MAGNET

A circular lifting magnet for a crane is to be designed so that with a flux density of 30,000 lines per square inch in each air gap, the length of each air gap is 0.5 inch. Leakage and saturation effects are such that the magnetomotive force for the air gaps is 0.85 of the magnetomotive force for the complete magnetic circuit. The mean length per turn of winding is 24 inches and the magnet is to operate with an applied terminal voltage of 70 volts with the winding at a temperature of 60°C.

(1) Wt. 2 What is the pull or tractive force in pounds per square inch for the magnet?

(2) Wt. 8 What size wire should be used for the winding?

SOLUTION

Assume a d-c magnet.

Force in dynes $= \frac{B^2A}{8\pi}$ Maxwell's Eqn

Force in lbs/sq. in. $= \frac{B^2}{72x10^6} = \frac{(30,000)^2}{72x10^6}$

(1) Force = 12.5 lbs/sq. in. pull

(2)

B = 30,000 Lines/sq. in., L = 0.5 in. air gap

$mmf_{air\ gap} = 0.85$ total NI

L_{mean} of coil = 24 inches

V = 70 volts on coil

$H_{air\ gap} = 0.313$ B NI per inch for air gap

$NI_{for\ air\ gap} = 0.313x30,000x0.5x2$ air gaps

= 9,400 Ampere turns for air gap

$NI_{total} = \frac{9,400}{0.85} = 11,050$ Ampere turns for air gap & magnetic ckt.

Since the coil dimensions are not given, assume 400 square inches of radiating surface and allow 0.7 watts per square inch dissipation at 60°C.

Watts = 400 x 0.7 = 280 watts dissipated

and $I = \frac{280 \text{ watts}}{70 \text{ volts}} = 4$ Amperes coil current

$N = \frac{NI}{I} = \frac{11050}{4} = 2763$ Turns on coil

Wire Length $= \frac{2763 \text{ x } 24 \text{ inches}}{12 \text{ inches/foot}} = 5526$ feet

$R_{wire} = \frac{70 \text{ volts}}{4 \text{ amperes}} = 17.5$ ohms

$R_{wire} = \frac{\rho \text{ length}}{\text{area}}$ where $\rho = 12$ at 60°C.

$A_{cir.\ mills} = \frac{12 \text{ x } 5526}{17.5} = 3880$ cir. mills

No. 14 AWG Magnet wire has 4,107 cir. mills and 2.525 ohms/1000'.

No. 15 AWG Magnet wire has 3,257 cir. mills and 3.184 ohms/1000'.

Choose No. 14 AWG Magnet wire.

ELECTRO-MAGNETIC FIELDS 3

FIELDS: POYNTING VECTOR

(1) Wt. 2 What is the Poynting vector? To what is it equal in terms of electric and magnetic quantities? Explain symbols or terms used carefully.

(2) Wt. 4 Consider the negative conductor of a two-wire, d-c, transmission line. Making free use of sketches, explain in words or demonstrate analytically how the Joulean energy loss is supplied by the Poynting vector. Indicate Poynting vector directions internal and external to the conductor.

(3) Wt. 4 Consider a 3-phase, 2-pole, 60-cps, squirrel cage induction motor the rotation of which is clockwise when viewed from the drive shaft end. Assume the motor is operating at full load and explain in words or demonstrate analytically how the energy made available in the stator circuit crosses or flows through the air gap of the machine by use of the Poynting vector. Free use of sketches should be made.

SOLUTION

(1) Poynting's Law states that transfer of energy can be expressed as the product of the values of magnetic field and the component of the electric field perpendicular to the magnetic field. The energy flow is in a direction perpendicular to both fields at any point.

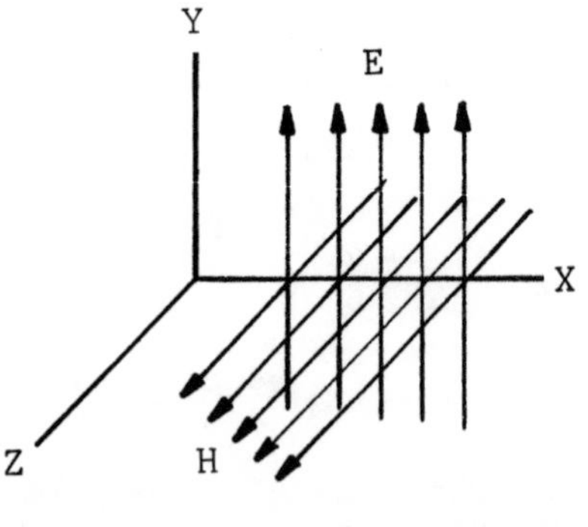

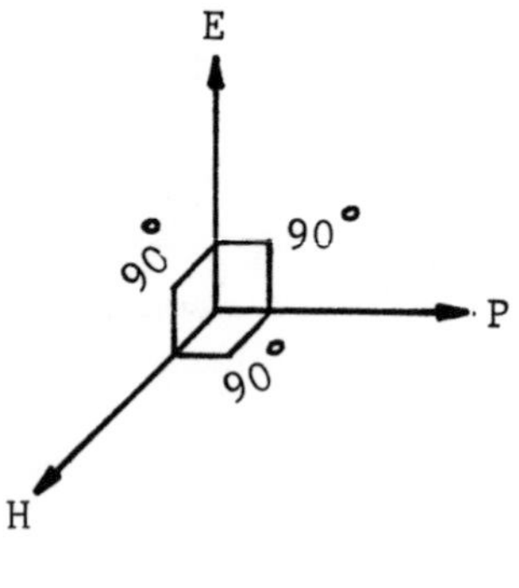

E is electric field intensity in volts per meter.

H is magnetic field in amperes per meter.

P is poynting vector in watts per square meter.

Energy flow is perpendicular to the E & H planes, i.e., the Y & Z planes, and is in watts per square meter. $\overline{P} = \overline{E} \times \overline{H}$ in vector form or $P = EH\sin\theta$ where θ is the angle between E & H.

(2) Consider an imaginary cylindrical surface about a conductor with a field "H" (due to the current) as shown and two components of an electric field "E", one component (up) representing the IR drop in the conductor, the other component represents the electric field between the lines. The poynting vector will then have a

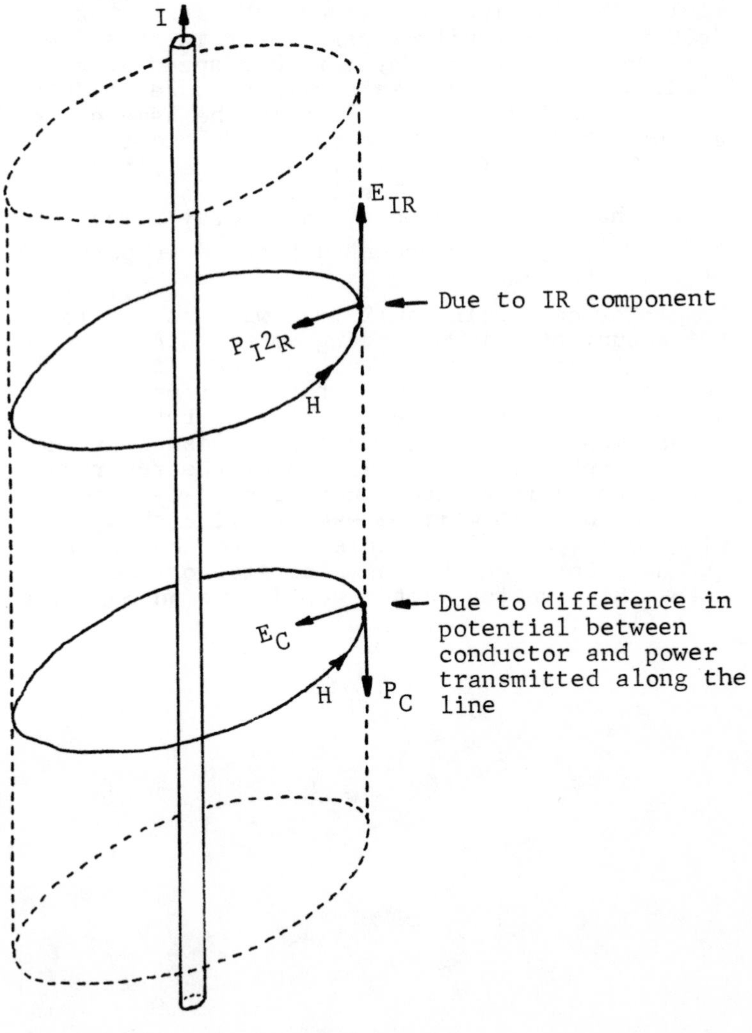

direction dependent on the electric field components, one component in towards the center of the conductor (representing the copper loss of the conductor), and the other in the direction of the conductor (representing the power transmitted to the load).

The poynting vector "$\overline{P}$" is then the vector sum of $\overline{P}_{I^2R} + \overline{P}_C = \overline{P}$.

(3) The 3Ø, 2 pole, 60 cycle squirrel-cage induction motor has 3Ø energy applied to the stator which produces a 2 pole 3 phase rotating field. The resultant flux vector of the 3Ø field travels clockwise around the stator at 3600 rpm. The squirrel-cage rotor in this field will rotate at almost synchronous speed at no load. The stator currents will produce an "H" field. The "E" field will be in the same plane as the "H" field. Perpendicular to the resultant "E" and "H" fields will be the resultant poynting vector "P". The instantaneous poynting vector has two components, the energy component and the I^2R loss component explained in part (2) of this problem.

The rotor delivering full load will have slip, the amount of slip depending on the N.E.M.A. class of motor. As a result of the slip, voltage will be induced in the rotor proportional to BLV. Rotor current will flow but in opposition to stator current as a result of the induced rotor voltage. If one of the rotor bars (conductors) is treated instantaneously, the poynting diagram will be exactly like the diagram in part (2) of this problem, on one side of the rotor. On the opposite side of the rotor the poynting vector would have an opposite direction.

ELECTRO-MAGNETIC FIELDS 4

A section of nonloaded telephone cable, located in a rural area, has the folowing characteristics per loop mile at a frequency of 1000 hertz:

Series resistance, 85.8 ohms
Inductance, 1.00 millihenry
Capacitance, 0.062 microfarad
Shunt conductance, 1.50 micromho

The cable consists of 400 pairs of No. 19 AWG, is shielded, and is 3.06" o.d.

REQUIRED:

Compute the following parameters for this cable at 1000 hertz:

Wt.		
3	(a)	characteristic impedance
3	(b)	attenuation in decibels per mile
2	(c)	phase shift per mile
2	(d)	velocity of propagation

Note: For full credit, you must show all of your work.
Answers taken directly from reference books are not acceptable.

SOLUTION

(a)

The characteristic impedance of a line, or Z_c is a complex expression composed of:

z = series impedance per unit length, per phase = $R + j \times 2\pi f L$
y = shunt admittance per unit length, per phase to neutral = $G + j2\pi f c$

Converting to henry, farad and mho (multiply by 10^{-3} or 10^{-6})

$$z = 85.8 + j\,2\pi \times 1000 \times 10^{-3} = 85.8 + j\,6.28 = 85.9 \angle 4^\circ \text{ ohm / mile}$$

$$y = (1.5 + j\,2\pi \times 1000 \times 0.062) \times 10^{-6} = (1.5 + j\,389) \times 10^{-6} =$$

$$= 389 \times 10^{-6} \angle 89.9^\circ \text{ mho / mile}$$

$$Z_c = \sqrt{\frac{z}{y}} = \sqrt{\frac{85.9 \angle 4^\circ}{389 \times 10^{-6} \angle 89.9^\circ}} = \sqrt{0.221} \times 10^6 \angle \frac{4^\circ - 89.9^\circ}{2} =$$

$$= 4.7 \times 10^2 = \angle -43^\circ = 470 \angle -43^\circ \text{ ohms ANS}$$

(b)

Let us denote a complex quantity γ as the propagation constant and l the line length in miles:

$$\gamma = \sqrt{z\,l\,y\,l} = \sqrt{zyl^2} = l\sqrt{zy}$$

The real part of the propagation constant γ is called "attenuation constant" (β) and is measured in nepers per unit length; the quadrature part of (γ) is called the "phase constant" (β) and is measured in radians per unit length; thus,

$$\gamma l = 1 \times \sqrt{yz} = 1.0 \times \sqrt{85.9 \times 389 \times 10^{-6}} \;\angle\frac{4^\circ + 89.9^\circ}{2}$$

$$= 0.182 \;\angle 47^\circ = 0.124 + j\,0.133 \text{ (radians)} = \alpha + j\beta$$

Converting nepers into decibels, we obtain:

α = 0.124 nepers / mile = 0.124 x 8.686 = 1.08 decibels/mile ANS.

(c) Converting radians into degrees: 180° = 3.1416 radians, then 0.133 radians = 7.6°, thus phase shift or β = 7.6°/mile ANS.

(d) A wavelength is the distance along a line between two points of a wave which differ in phase by 360° or 2π radians. If λ is the phase shift in radians per mile, the wavelength in miles is

$$\lambda = \frac{2\pi}{\beta} = \frac{6.28}{0.133} = 47.2 \text{ miles}$$

Velocity of propagation is the product of the wavelength in miles and the frequency in cycles per second, or

velocity = λf = 47.2 x 1000 = 47,200 miles / second ANS.

Reference: Elements of Power System Analysis, by William D. Stevenson, Jr., McGraw Hill, p. 100-106.

ELECTRO-MAGNETIC FIELDS 5

Shown below is a diagram of a coaxial transmission line in which the circuit elements have the values indicated, and E_g is the open circuit generator voltage, Z_o is the characteristic impedance of all transmission lines, and L_2 is greater than one wavelength.

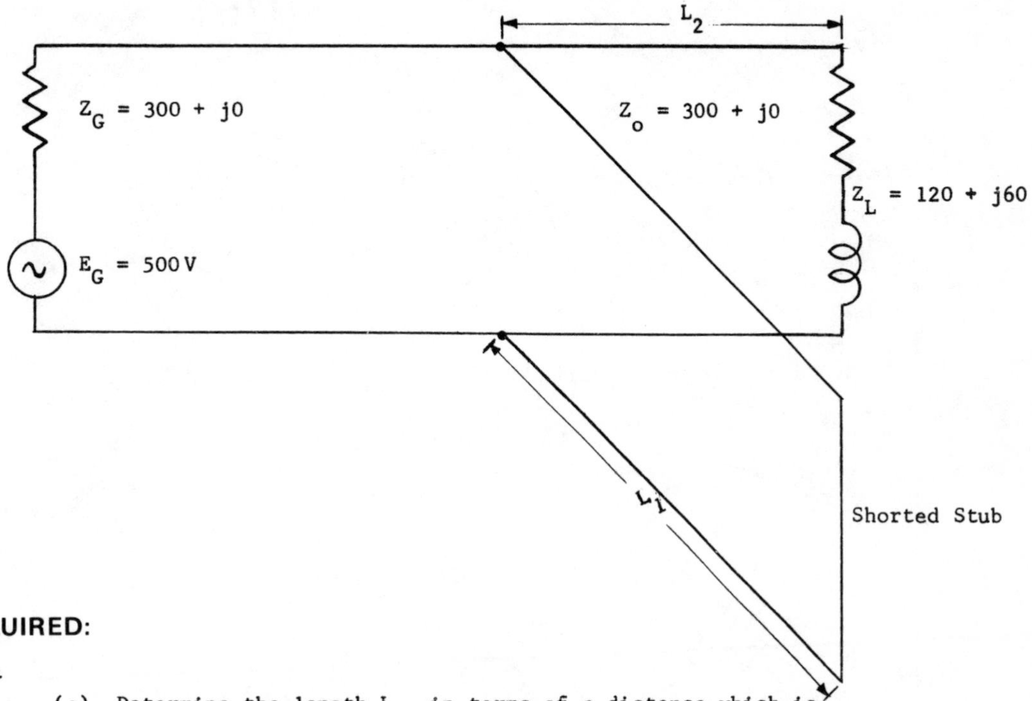

REQUIRED:

Wt.		
3	(a)	Determine the length L_2, in terms of a distance which is greater than one wavelength, and is the proper distance so that a shorted stub attached at this point will cause the line to be matched to Z_o.
3	(b)	Determine the length of the shorted stub L_1 needed to match the load to the line at this point.
2	(c)	Determine the VSWR on the unmatched portion of the line L_2.
1	(d)	Determine the power that will be supplied to the load under the matched condition.
1	(e)	Determine the greatest voltage that will appear across the transmission line in the unmatched section L_2.

NOTE: A Smith Chart is included to aid in the solution, or a straight analytical solution may be employed.

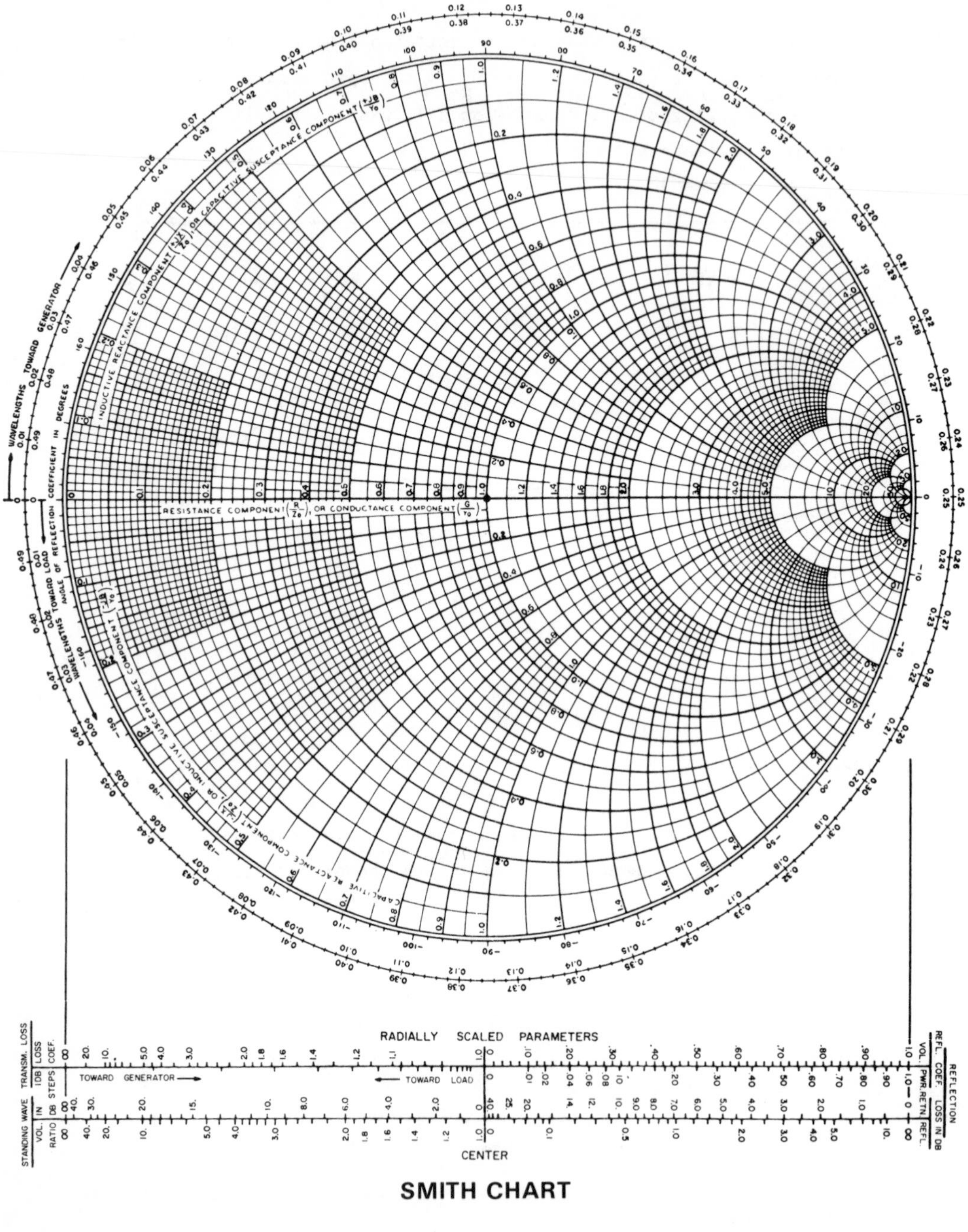
RESISTANCE COMPONENT (R/Zo), OR CONDUCTANCE COMPONENT (G/Yo)
INDUCTIVE REACTANCE COMPONENT (+jX/Zo), OR CAPACITIVE SUSCEPTANCE COMPONENT (+jB/Yo)
CAPACITIVE REACTANCE COMPONENT (-jX/Zo), OR INDUCTIVE SUSCEPTANCE COMPONENT (-jB/Yo)
WAVELENGTHS TOWARD GENERATOR
WAVELENGTHS TOWARD LOAD
ANGLE OF REFLECTION COEFFICIENT IN DEGREES
RADIALLY SCALED PARAMETERS
TOWARD GENERATOR
TOWARD LOAD
CENTER
STANDING WAVE
VOL. RATIO
IN DB
TRANSM. LOSS
1DB STEPS
LOSS COEF.
REFLECTION
REFL. COEF.
VOL.
PWR.
RETN.
LOSS IN DB
REFL.

SMITH CHART

SOLUTION

As a first step, the Y_γ/Y_o ratio has to be determined:

$$Y_\gamma = \frac{1}{Z_\gamma} = \frac{1}{120 + j60} = \frac{1}{120 + j60} \times \frac{120 - j60}{120 - j60} = 0.0067 - j0.0033$$

$$Y_o = \frac{1}{Z_o} = \frac{1}{300} = 0.0033$$

$$\frac{Y}{Y_o} = \frac{0.0067 - j0.0033}{0.0033} = 2 - j1$$

This ratio is point A on the Smith Chart.

(a) To determine the length L_2 the stub is placed at the point where $G = 1/Z_o$ or at $G = 1$ which is point B on the Smith Chart. However, L_2 must be larger than one wavelength, so that location is 360° further away from load than indicated.

Initial location:	-26.5°
Final location:	-61.8°
Difference	35.3°

$$\frac{35.30°}{2} = 17.65°$$

The proper distance of the stub: 17.65° + 360° = 377.65° ANS.

(b) The length of the shorted stub L_1 is such that reactance of type opposite to line is required.

On Smith Chart start at U_1 ($Y_{\gamma\,stub}/Y_o = \infty$), move CCW to U_2 where the imaginary component of $G + jB$ is opposite that of stub location but equal in magnitude.

Initial location:	0°
Final location:	270°
Difference	270°

The length of the stub: $\frac{270°}{2} = 135°$ ANS.

(c) To determine the VSWR on the unmatched portion of the line L_2, from point C on Smith Chart, we obtain

$N = 2.6$ ANS.

This value can also be obtained as follows:

$$\rho = \frac{Z_L - Z_o}{Z_L + Z_o} = \frac{120 + 6j0 - 300 - j0}{120 + j60 + 300 + j0} = \frac{-180 + j60}{420 + j60} \times \frac{420 - j60}{420 - j60} = 0.4 + j0.2$$

$$|\rho| = \sqrt{0.4^2 + 0.2^2} = 0.446$$

$$\text{VSWR} = N = \frac{1 + |\rho|}{1 - |\rho|} = \frac{1 + 0.446}{1 - 0.446} = \frac{1.446}{0.554} = 2.6 \text{ q.e.d}$$

(d)

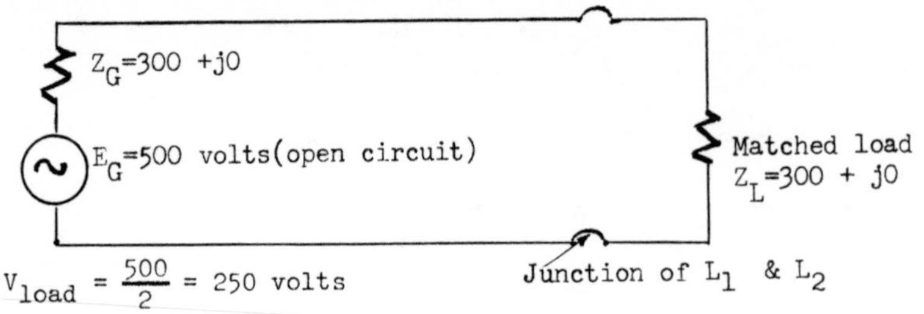

$V_{load} = \frac{500}{2} = 250$ volts

The power then will be supplied to the load under the matched condition:

$P_{load} = \frac{E^2}{R} = \frac{(250)^2}{300} = 208$ watts ANS.

(e) $V_{max} = V_{load}\,(1 + |\rho|) = 250 \times 1.446 = 361$ volts ANS.

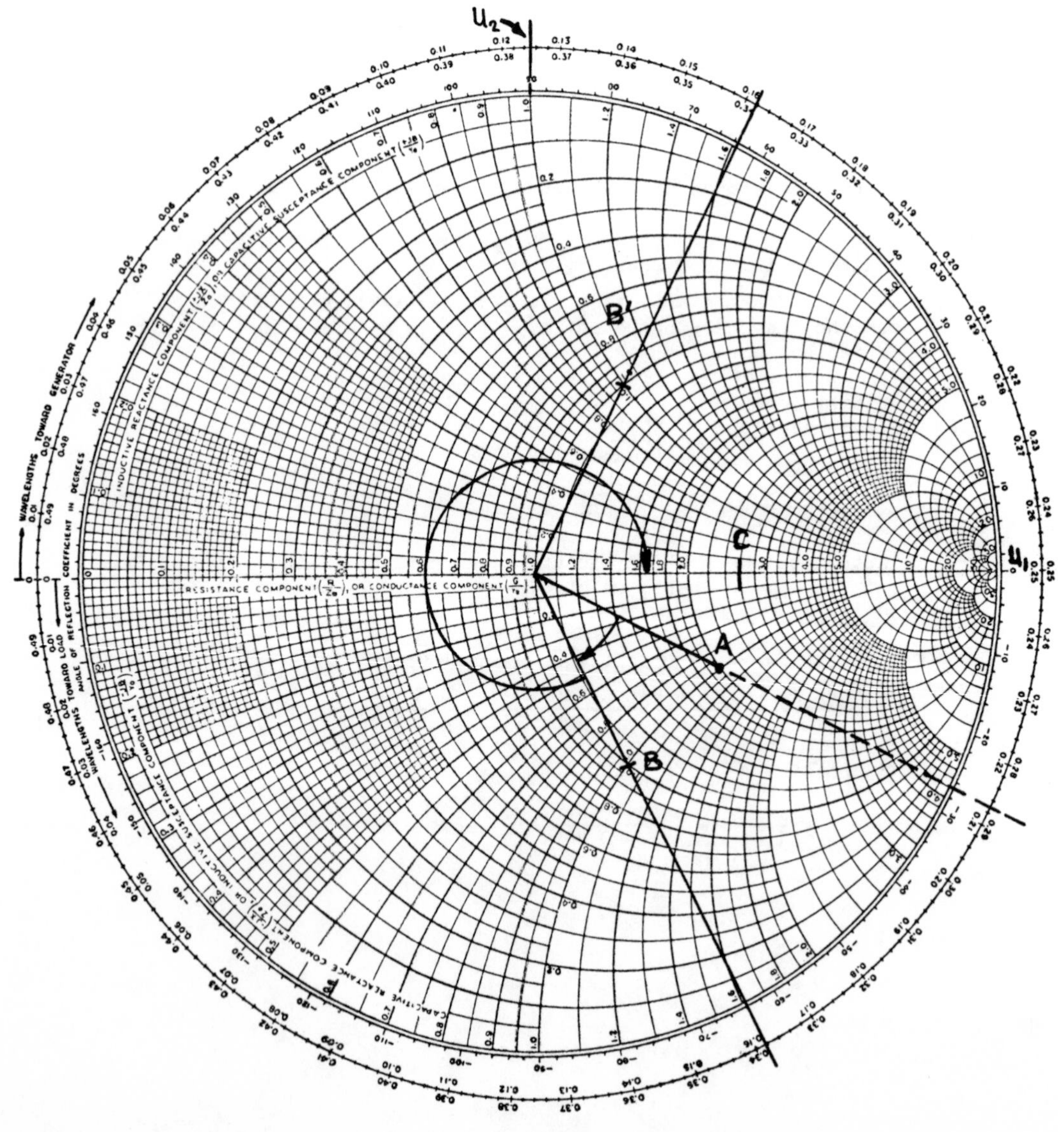

4

Machinery

The examinations in the past have been relatively strong in the machinery and power distribution areas (refer to the appendix). To solve most of the problems in this area one not only needs to review d.c. and a.c. machinery, but also the basic equivalent circuit of a transformer. This equivalent circuit gives the foundation for computing many other aspects of a.c. machinery.

Generally, the sequence of review should follow the study format of

1. d.c. machine theory
2. transformer equivalent circuit theory (and three phase transformer connections)
3. induction motor theory
4. synchronous machine theory.

In the author's opinion, if one has trouble with the basic transformer equivalent circuit calculations and theory (or never has studied it in the past), he probably should not spend time reviewing the rest of a.c. machinery, but instead should use the study time in the areas of circuits, control systems, or electronics.

D.C. MACHINE AREA

This material is usually straight forward and will not be reviewed here except to emphasize a few basic relationships. The equivalent circuit is as follows:

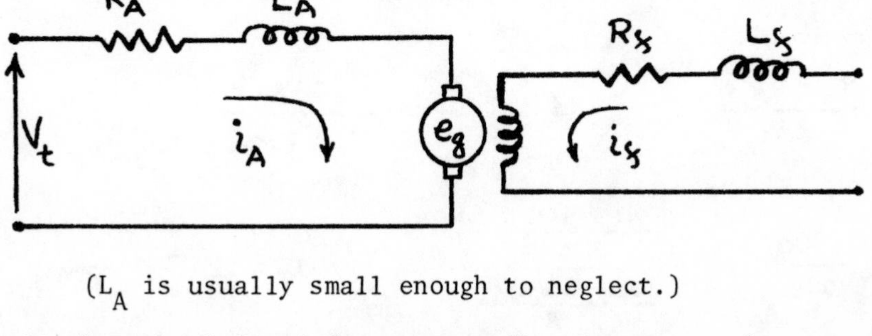

(L_A is usually small enough to neglect.)

Recall that the induced (or generated) voltage, e_g, is proportional to both speed and magnetic flux (usually considered to be linearly proportional to the field current). Then, from Kirchhoff's law, the resulting equation is:

$$V_t = R_A i_A + L\frac{di_A}{dt} + e_g \cong R_A i_A + e_g$$

and e_g is proportional to i_g and speed ω.
Then, if one solves for the speed, the equation becomes:

$$\omega = \frac{V_t - i_A R_A}{K i_g}$$

And, from this equation, many other equations may easily be derived. As an example, if ω = 0 (on starting), the starting current becomes

$$i_A\,(\text{start}) = V/(\Sigma R)$$

where $\Sigma R = R_A + R_{start}$, and so on.

If, however, a problem should ask for transient conditions, one would usually be better advised to use Laplace transforms and use the methods of analysis suggested in the chapter on Control Systems.

For steady state conditions, for example, one could find new operating characteristics for new field settings based upon original data. Thus if original data was given as 1800 rpm with a field current of 2 amperes (with a known R_A of 0.1 ohm and an armature rated current of 50 amperes and a rated voltage of 200 volts), a question might be to find the new speed if the field current was changed to 3 amperes with an armature current of 1/2 of rated. To solve this problem, one only needs to take a ratio of the previous speed equation for the original conditions and new conditions:

$$\omega_1 = \frac{V_t - i_{A_1} R_A}{K i_{g_1}} \quad \text{and} \quad \omega_2 = \frac{V_t - i_{A_2} R_A}{K i_{g_2}}$$

$$\therefore \frac{1800}{\omega_2} = \frac{\left(\frac{200 - 50 \times 0.1}{K2}\right)}{\left(\frac{200 - 25 \times 0.1}{K3}\right)}, \quad \omega_2 = 1215.4 \text{ rpm}$$

TRANSFORMER AREA

Most problems have dealt with transformers using steady state theory for an equivalent circuit at a particular frequency. Recall that an ideal transformer has the voltage referred to one side directly as the turns ratio, the current inversely, and the impedance reflected through as the square of the turns ratio.

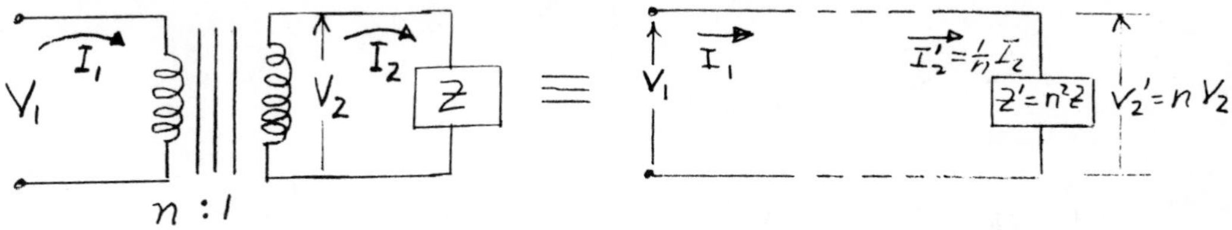

Then for the non-ideal tranformer, the approximate relationship may be given as:

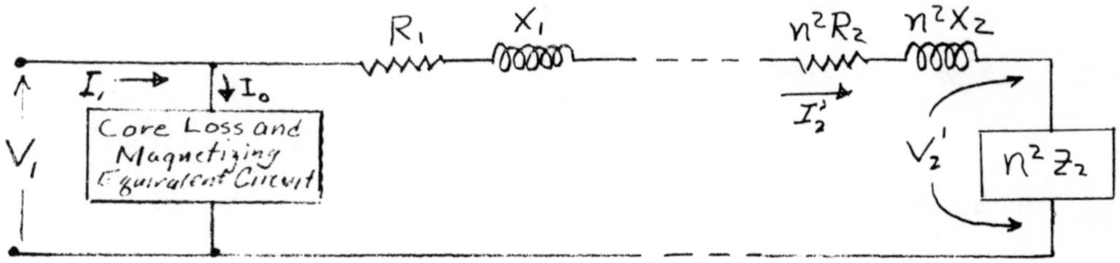

Then $\bar{I}_1 = \bar{I}_0 + \bar{I}_2'$, here (and in almost all phasor problems involving a.c. steady state circuit theory) it is a good idea to actually lay out to scale the phasor diagram (bring a ruler and protractor to the exam along with your calculator). As an example, if one wanted to find the actual V_1 necessary to supply the correct load voltage, V_2, for some particular load, Z, then merely find $\bar{I}_2' = \frac{\bar{V}_2'}{Z_L'} = \frac{n\bar{V}_2}{n^2 Z_L} = \bar{I}_2' \angle\theta$

Since the equivalent circuit, as far as the source V_1 is concerned, is a series one, add all voltage drops (by phasors) to yield V_1:

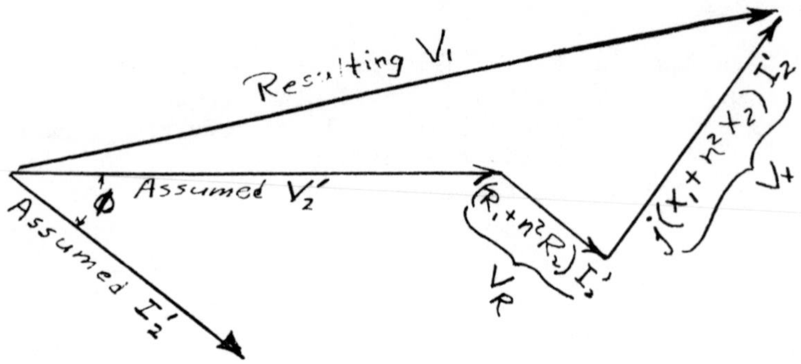

(Of course V_R is parallel and V_X is perpendicular to $\bar{I}_2'$.)

The efficiency of the device transformer can easily be found with the relationship:

$$\text{Eff} = \frac{\text{Power to Load } (Z_L')}{\text{Power to Load} + \text{All losses}}$$

where losses are core and $(I_2')^2(R_1 + n^2R_2)$

SYNCHRONOUS MACHINES

To solve problems involving three phase synchronous machines one needs to recall that the key quantity for most of these problems is the synchronous reactance, X_s. (Usually X_s is large enough to allow one to neglect both the internal resistance and the leakage reactance.) This synchronous reactance, although a nonlinear function of the d.c. field current, is fairly constant and is considered so for most of the exam type problems.

Assuming that X_s is constant, the easy way to solve most of these problems is to assume the equivalent circuit is one leg of a wye with a returned neutral and to work on a per phase basis:

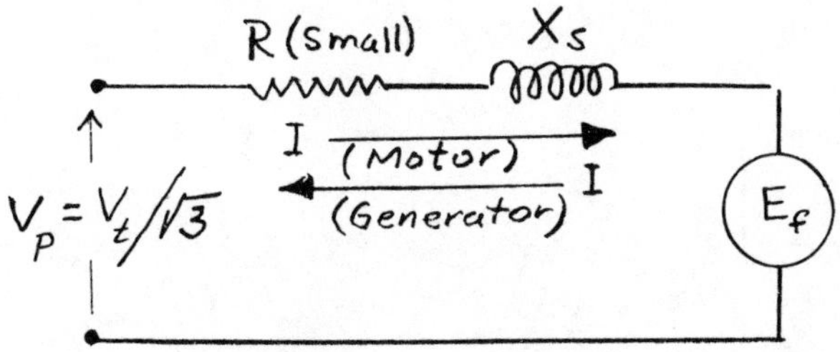

E_f = Generated voltage due to the dc field current (frequently considered to be linearly related).

V_t = Rated terminal voltage

V_p = Phase terminal voltage

(Again, in solving these problems, one is urged to use a ruler and protractor and lay out the phasor diagram to scale.)

To work a synchronous generator (or motor) problem, one usually considers the terminal voltage as the known quantity, (that is, name plate value) and with the KVA rating (KVA = $\sqrt{3}$ VI/1000) of the machine, one can begin to set up the problem solution from the equivalent circuit. As an example, one may operate at a leading or lagging power factor by varying the d.c. field current (i.e., for a motor, lowering the field tends towards a lagging P.F., while increasing the field tends towards a leading P.F.). And, from the equivalent circuit for any condition of leading or lagging P.F. for a particular current, one then can calculate the voltage, E_f, and in turn, the field current needed (from a voltage vs excitation curve) as follows for a generator:

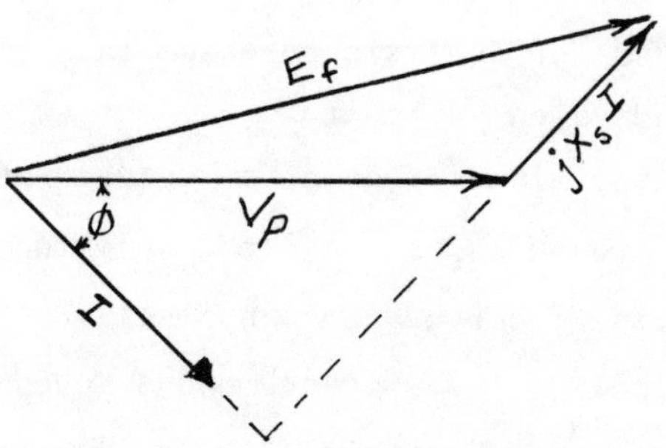

Lagging P.F.

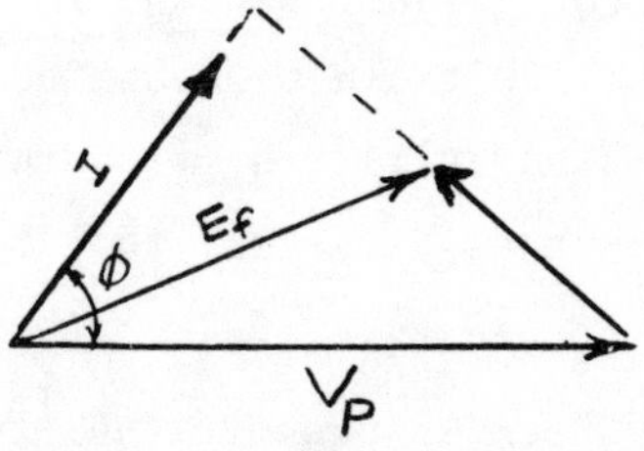

Leading P.F.

And for a motor (along with the "V" loading curves):

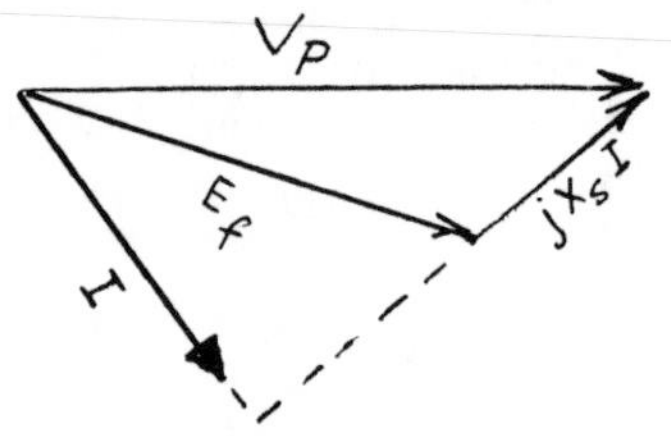

Lagging P.F.
(Under-excited)

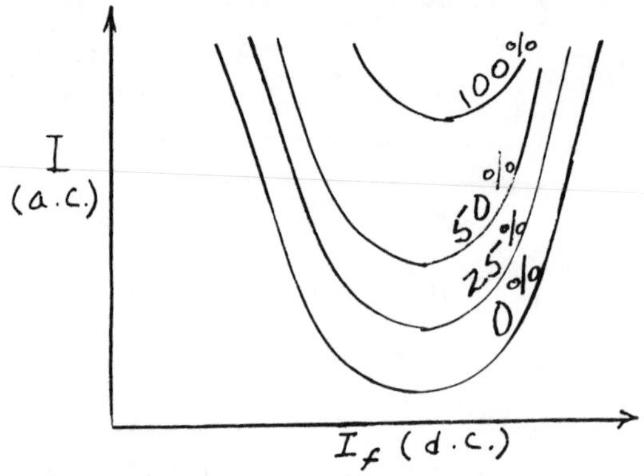

Percent of Full-Load "V" Curves

Of course, torque can always be found from the relationship:

$$T = \frac{P_{total}(\text{watts})}{\omega\ (\text{rad/sec})} \quad \text{Newton meters}$$

Here the problem of "bookkeeping" is important as one is involved with 1/3 values for the equivalent circuit, and full values for the overall machine.

INDUCTION MOTORS

For a squirrel cage induction motor, the transformer equivalent circuit can be modified and used for analysis purposes (at standstill, the transformer equivalent circuit is correct as it stands). Recall that the machine always runs at an actual speed, n_a, that is less than the synchronous speed, n_s (the synchronous speed is proportional to the line frequency and inversely proportional to the number of poles). This synchronous speed is the speed of the rotating flux vector; and the slip, s, is the difference between the synchronous and actual speed, divided by the synchronous speed. The transformer circuit modification then accounts for the power developed within the machine; this may be accounted for by a fictitious load resistor R_L so that I^2R_L represents 1/3 of the total power developed. (Here, one has to be careful to account for the friction and windage loss; this "bookkeeping" loss may either be lumped with the core loss or subtracted from the developed power.) The modified equivalent circuit now becomes:

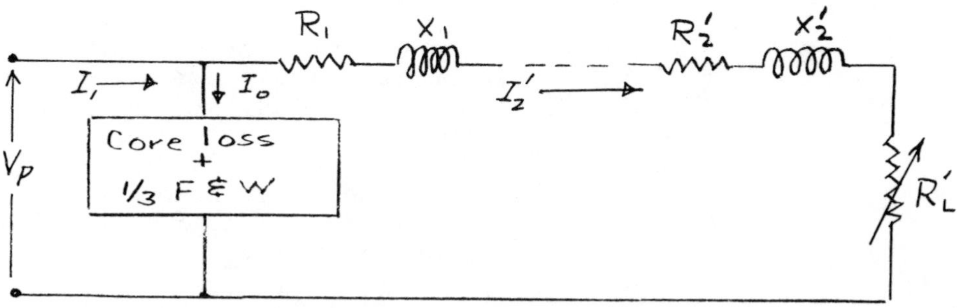

where $R_L' = R_2' \left(\frac{1-S}{S}\right)$

Thus for many problems, if one knows the speed (and therefore the slip), then the fictitious load R_L' may be calculated; and from this, the developed power is $3\,I_2'^2 R_L'$ and the developed torque is found from $T = \frac{P}{\omega}$ (again, if the friction is not included in core loss, it must be substracted from the developed power).

The parameters $R_E = R_1 + n^2R_2$ and $X_E = X_1 + n^2X_2$ may be found from the "short circuit" condition (blocked rotor) and the core loss + F&W may be found from the "open circuit" condition (running at no load).

Of course, all of the induction motors run at a lagging power factor, and in a good many of the exam problems, the key is (especially if two or more machines are paralleled on the same lines) one of "bookkeeping" with all additions being that of phasor addition. Again, carefully lay out the problem and draw the phasor diagrams (let the math follow from the diagrams rather than the other way around).

MACHINERY 1

(1) <u>Wt. 2</u> A three phase induction motor has the following data on its name plate: 10 hp, 440 volts, 14.1 amps., 60 cps, 1750 rpm. If the motor is operated from a 370-volt, 3-phase, 50-cycle-per-second source, the rotor speed when delivering its name plate rated torque will be approximately:

a. - 1750 rpm	g. - 1460 rpm
b. - 1700 rpm	h. - 1420 rpm
c. - 1660 rpm	i. - 1380 rpm
d. - 1600 rpm	j. - 1340 rpm
e. - 1560 rpm	k. - 1300 rpm
f. - 1500 rpm	l. - None of these

(2) <u>Wt. 2</u> A 5-hp, 220-volt, d-c shunt motor draws 21.0 amperes at rated load and speed. Manipulating the field control rheostat so as to increase the shunt field circuit resistance by approximately 10% will:

a. - Increase the field circuit current
b. - Decrease the line voltage
c. - Increase the line voltage
d. - Decrease the line current
e. - Decrease the speed
f. - Increase the speed
g. - Cause line current to have lagging power factor
h. - Decrease torque supplied
i. - None of these

(3) Wt. 2 An alternator is operating at 1.0 power factor in parallel with a large system. Increasing the field current will:

a. - Decrease the speed
b. - Increase the speed
c. - Increase the line voltage
d. - Decrease the line voltage
e. - Decrease the line current
f. - Increase the line current
g. - Increase the torque supplied
h. - Decrease the torque supplied
i. - None of these

(4) Wt. 2 A 5 kva, 2400 - 120/240 volt distribution transformer when given a short-circuit test had 94.2 volts applied with rated current flowing in the short-circuited winding. The per unit impedance of the transformer is approximately:

a. - 0.02
b. - 0.025
c. - 0.03
d. - 0.035
e. - 0.04
f. - 0.045
g. - 0.05
h. - 0.055
i. - 0.06
j. - None of these

(5) Wt. 2 In starting a 500 horsepower, 2300 volt, three phase synchronous motor, the field winding is initially short-circuited so as to:

a. - Produce much larger starting torque
b. - Lower induced voltage in field winding
c. - Increase induced voltage in field winding
d. - Provide better flux distribution in the air gap
e. - Lower voltage produced between layers of the field windings
f. - Raise voltage produced between layers of the field windings
g. - Shorten acceleration time
h. - Increase acceleration time
i. - None of these

SOLUTION

(1) g. - 1460 rpm approximately.

$$N_{syn} = \frac{f_1 \times 2 \times 60}{p} = \frac{60 \times 2 \times 60}{4} = 1800 \text{ rpm syn. speed}$$

Slip = 1800 - 1750 = 50 rpm

$$s = \frac{1800 - 1750}{1800} = \frac{1}{36}$$

machine has 4 poles

$$N_{syn} \text{ on 50 cps} = \frac{50 \times 2 \times 60}{4} = 1500 \text{ rpm}$$

$$N_{rotor} = 1500 - \frac{1}{36} \times 1500 = 1458.0 \text{ rpm on 50 cps}$$

(2) f. - Increase armature speed.

$$N_{rpm} = \frac{V_T - I_a R_a}{K\phi_f} \quad \text{where } \phi_f \propto I_f$$

If R_f is increased I_f decreases as does ϕ_f.

so N_{rpm} increases.

(3) f.

Increasing field current will increase armature current. Internal torque demand increases momentarily which will tend to increase load on prime mover. If the prime mover has a governor controlling energy delivered to it, more power will increase line current.

Alternators operating in parallel must have the same frequency and terminal voltage. Increasing the field current will slightly increase the line current which in turn will put a greater load on the prime mover, tending to slow it down. But being locked into the system, the motor-generator cannot slow down, so more power would have to be delivered to the prime mover in order to generate more power. The load delivered by alternators in parallel cannot be changed appreciably by means of the alternator fields. Loads on alternators operating in parallel are changed by shifting the speed-load characteristics of the prime mover. That is, either increasing or decreasing the prime mover power by more or less fuel.

(4) e. - Z per unit = 0.04 approximately.

$$I_H = \frac{5000}{2400} = 2.08 \text{ Amperes rated high voltage coil current}$$

$$I_L = \frac{5000}{240} = 20.8 \text{ Amperes rated low voltage coil current}$$

$$Z_H = \frac{94.2}{2.08} = 45.3 \text{ ohms}$$

$$E_{base} = 2400$$

$$I_{base} = 2.08 \text{ Amperes}$$

$$Z_{base} = \frac{2400 \text{ volts}}{2.08 \text{ amperes}} = 1152 \text{ ohms}$$

$$Z/\text{unit} = \frac{Z_H}{Z_{base}} = \frac{45.3}{1152} = 0.0392 \text{ ohms}$$

(5) **e. - Lower voltage produced between layers of the field windings. Also, b. would apply.**

The field winding of large synchronous motors has many turns and a large amount of inductance. When voltage is first applied to the stator, high voltages are induced in in the dc rotor field windings if the circuit is open. If the field winding is shorted, short circuit current will flow. The field terminal voltage will be zero and the insulation will not be punctured. If the field is left open, a very high open circuit terminal voltage will develop which may puncture the turn-to-turn insulation, ruining the field winding.

MACHINERY 2

POWER MACHINERY: SHUNT MOTOR

A d-c shunt motor has a name plate rating of 15 hp, 230 volts, 57.1 amp, 1400 rpm. The field circuit has a resistance of 115 ohms and the armature circuit resistance is 0.13 ohm. Neglecting the effect of armature reaction, find:

(1) Wt. 6 The no-load line current

(2) Wt. 4 The no-load speed

SOLUTION

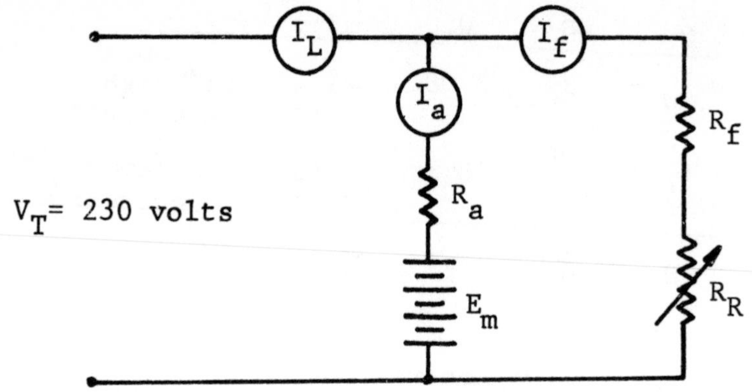

Neglect Armature reaction

$I_a = I_L - I_f = 57.1 - 2 = 55.1$

Input $= V_T I_L = 230 \times 57.1 =$	13,200	watts
$V_T I_f = 230 \times 2 =$	460	"
Armature Input =	12,740	watts
$I_a^2 R_a$ loss $= (55.1)^2(0.13) =$	396	"
	12,344	watts
Power output = 15 x 746 =	11,200	"
Rotational Losses =	1,144	watts

No Load Losses $= V_T I_a + I_a^2 R_a$

$230\ I_a - I_a^2(0.13) = 1144$

$0.13\ I_a^2 - 230\ I_a + 1144 = 0 \qquad ax^2+bx+c=0$

$$x = \frac{-b \pm \sqrt{b^2-4ac}}{2a}$$

$$I_a = \frac{230 \pm \sqrt{(-230)^2-4(0.13)(1144)}}{2(0.13)}$$

$$= \frac{230 \pm \sqrt{53200 - 595}}{0.26} = \frac{230 \pm \sqrt{52,605}}{0.26} = \frac{230 \pm 229}{0.26}$$

= 3.85 Amps or 1070 Amps

(1) No-load line current: I_a = 3.85 Amps no load

$I_L = I_a + I_f = 3.85 + 2.0 = 5.85$ amps

(2) No-load speed

$$N_{Arm} = \frac{V_T - I_a R_a}{K\phi_f} \quad \text{neglecting armature reaction}$$

$$N_{no\ load} = \frac{230 - 3.85 \times 0.13}{K\phi_f}$$

$$N_{full\ load} = \frac{230 - 55.1 \times 0.13}{K\phi_f} = 1400 \text{ rpm}$$

$$N_{no\ load} = \frac{230 - 0.53}{230 - 7.2} \times 1400 = \underline{1440 \text{ rpm}}$$

MACHINERY 3

A 50 KVA, 2300/230 volt, 60 cycle transformer is tested in the laboratory so that its characteristics may be determined. The standard test requires an open circuit test and a short circuit test.

Open Circuit Test - Core Loss

I	E	W
6.5	230	187

Short Circuit Test - Cu Loss

I	E	W
21.7	115	570

REQUIRED:

(a) Calculate the resistance of the windings.

(b) Calculate the copper loss.

(c) Calculate the core loss.

(d) Determine the efficiency of the transformer at full load.

(e) Determine the efficiency of the transformer at half load.

(f) Find the regulation of the transformer for power factor of 1.0.

(g) Find the regulation of the transformer for power factor of 0.8 lag.

(h) Find the regulation of the transformer for power factor of 0.8 lead.

SOLUTION

The open circuit test measures core loss with negligible copper loss. The short circuit test measures the copper loss with negligible core loss.

The coils are designed so that

$$I_1^2 R_1 = I_2^2 R_2$$

where I_1 and R_1 are the high voltage coil current and a.c. resistance.

I_2 and R_2 are the low voltage coil current and resistance.

$$\frac{E_1}{E_2} = \frac{N_1}{N_2} = \frac{I_2}{I_1}$$

where $\frac{N_1}{N_2} = a$ (turns ratio)

a) Since

$$W_{cu\ loss} = I_1^2 R_1 + I_2^2 R_2 = 570 \text{ watts}$$

and

$$I_1^2 R_1 = I_2^2 R_2 \text{ for good transformer design}$$

$$570 = 2I_1^2 R_1 = 2I_2^2 R_2$$

$$I_1 = \frac{50,000}{2300} = 21.7 \text{ amps (Rated)}$$

$$I_2 = \frac{50,000}{230} = 217 \text{ amps (Rated)}$$

$$\therefore\ R_1 = \frac{570}{2(21.7)^2} = 0.605\ \Omega \quad \text{(High Side)}$$

$$R_2 = \frac{570}{2(217)^2} = 0.00605\ \Omega \quad \text{(Low Side)}$$

b) Copper loss = $I_1^2 R_1 + I_2^2 R_2 = 570$ watts

c) Core loss = 187 watts neglecting the no load exciting current copper loss which would amount to $(6.5)^2(0.00605) = 0.255$ watt

d) Full load efficiency

$$= \frac{\text{OUTPUT}}{\text{output + Cu loss + CORE loss}}$$

assume Pf = 1

$$\text{Efficiency} = \frac{50{,}000}{50{,}000 + 570 + 187} = \frac{50{,}000}{50{,}757} = 0.985$$

e) Half load efficiency

assume Pf = 1

$$\text{Efficiency} = \frac{25{,}000}{25{,}000 + \frac{570}{4} + 187} = \frac{25{,}000}{25{,}329.5} = 0.99$$

f) Voltage regulation

$$= \frac{\text{No Load Voltage} - \text{Full Load Voltage}}{\text{Full Load Voltage}}$$

Convert the equivalent transformer circuit, referring it to the low voltage coil. Assume the low voltage coil has constant voltage $E_2 = 230$ volts

$$R_{Equiv.\,Low} = \frac{Power}{I_2^2} = \frac{570}{(217)^2} = 0.0121 \text{ ohm}$$

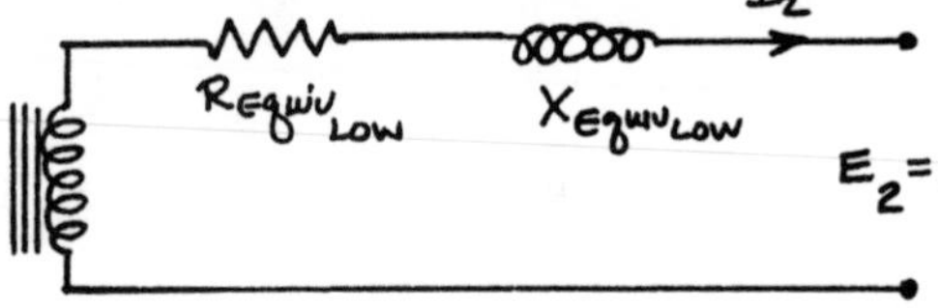

$$Z_{Equiv.\,High} = \frac{115}{21.7} = 5.3\ \Omega$$

$$Z_{Equiv.\,Low} = \frac{Z_{Eq.\,High}}{a^2} = \frac{5.3}{100} = 0.053\ \Omega$$

$$X_{Equiv.\,Low} = \sqrt{Z^2_{Equiv.\,Low} - R^2_{Equiv.\,Low}} = \sqrt{(0.053)^2 - (0.0121)^2}$$

$$= 0.0516 \text{ ohms}$$

Assume rated current $I_2 = 217$ amps

$$\frac{\bar{E}_1}{a} = E_2 + I_2 (R_{equiv.\,Low} + jX_{equiv.\,Low})$$

$$= 230 + 217(0.0121 + j\,0.053)$$

$$= 230 + 2.63 + j\,11.5$$

$$= 232.63 + j\,11.5$$

$$= 232.7 \angle 2.84°$$

Voltage Regulation at P.f. = 1 $= \dfrac{232.7 - 230}{230} = \dfrac{2.70}{230}$

$= 0.01175$ or 1.175 percent

g) Find Voltage regulation for P.f. = 0.8 Lag
Assume rated current $I_2 = 217$ amps

Draw Phasor Diagram

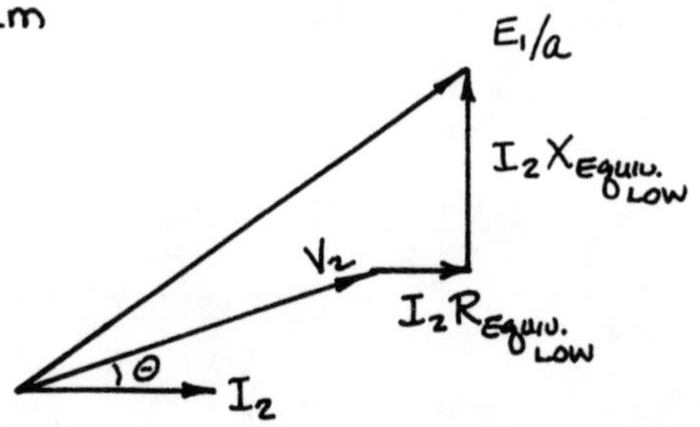

$$\frac{\bar{E}_1}{a} = V_2(\cos\theta + j\sin\theta) + I_2 R_{equiv_{LOW}} + j I_2 X_{equiv_{LOW}}$$

$$= 230(0.8 + j0.6) + 217(0.0121 + j0.053)$$

$$= 184 + j138 + 2.63 + j11.5$$

$$= 186.63 + j149.5$$

$$= 238.2 \angle 38.7°$$

$$\text{Voltage regulation} = \frac{238.2 - 230}{230} = \frac{8.2}{230} = 0.0356 \text{ or } 3.56\%$$

h) Voltage Regulation for P.f. = 0.8 Lead

Assume I_2 = 217 amperes, rated current

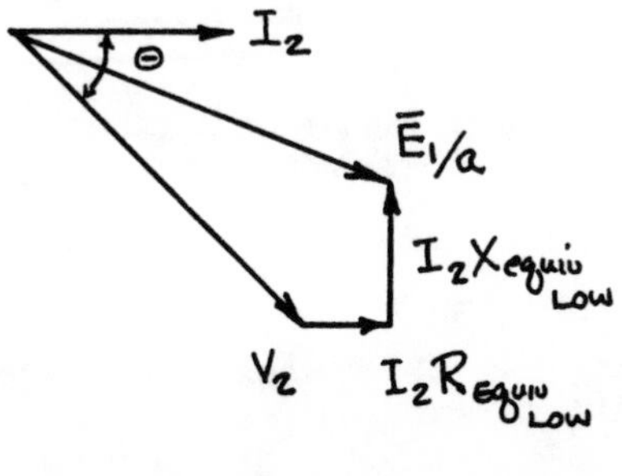

$$\frac{\bar{E}_1}{a} = V_2(\cos\theta - j\sin\theta) + I_2 R_{equiv_{LOW}} + j I_2 X_{equiv_{LOW}}$$

$$= 230(0.8 - j0.6) + 2.63 + j11.5$$

$$= 184 - j138 + 2.63 + j11.5$$

$$= 186.63 - j126.5$$

$$= 224 \angle 34.2°$$

$$\text{Voltage regulation} = \frac{224 - 230}{230} = -0.0262 \text{ or } -2.62\%$$

MACHINERY 4

POWER MACHINERY: SYNCHRONOUS MOTOR TORQUE

A 3-phase, Y-connected, 2300-volt, 50-cps 30-pole, 1000-hp, 1.0-power factor synchronous motor has a synchronous reactance of 3.7 ohms per phase. The motor is supplied from a 3-phase, Y-connected, 1000-kva, 2-pole, 3000-rpm turbine generator whose synchronous reactance is 5.1 ohms per phase.

The field currents of the motor and generator are maintained at those magnitudes which produce rated voltage with 1.0 power factor of the motor at rated motor load. With no other load supplied by the generator the motor mechanical load is gradually increased. Neglecting losses, what is the maximum motor torque in pound-feet that could be produced?

SOLUTION

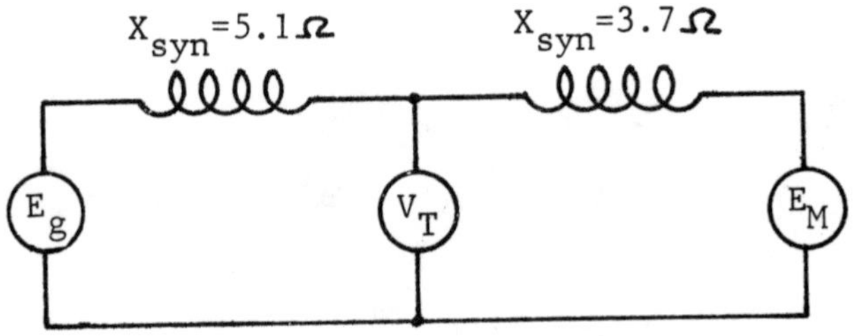

The problem will be solved on a per phase basis. The diagrams are per phase.

$$I_M = \frac{1000 \times 746}{2300\sqrt{3}} = 187 \text{ Amperes Rated Current}$$

$$V_T = \frac{2300}{\sqrt{3}} = 1325 \text{ Volts per phase}$$

$$E_M = V_T - jIX_s = 1325 - j187 \times 3.7 = 1500\angle{-26.5}$$

$$E_g = V_T + IX_s = 1325 + j187 \times 5.1 = 1630\angle{43.6}$$

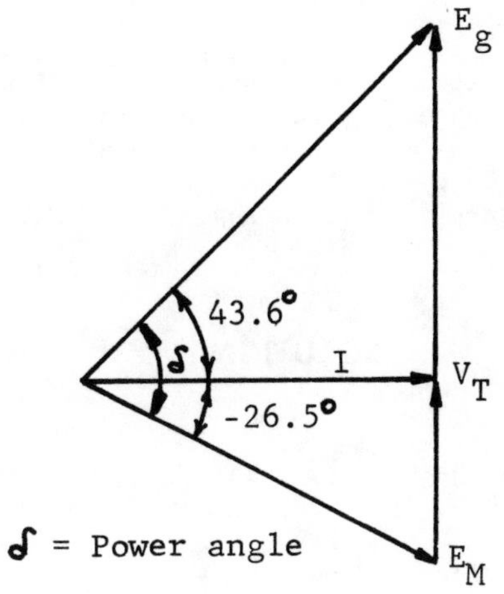

Under the conditions stated in the problem it can be shown that:

$$\text{Power}_{\text{maximum}} = \frac{E_g \cdot E_M}{X_s + X_s}$$

when the power angle equals 90°
(See: Carr, "Electrical Machinery", John Wiley and Sons, or
Fitzgerald & Kinsley, "Electric Machinery")

$$\text{Power}_{\text{max}} = \frac{1630 \text{ x } 1500}{5.1 \text{ x } 3.7} = 278{,}000 \text{ Watts/phase}$$

$$3 \text{ Phase Power}_{\text{max}} = 834{,}000 \text{ Watts}$$

$$\text{Torque}_{\text{max}} = \frac{7.04 \text{ x Poles}}{120 \text{ f}} \text{ x Watts}$$

$$= \frac{7.04 \text{ x } 30}{120 \text{ x } 50} \text{ x } 834{,}000 = 29{,}300 \text{ Ft-Lbs.}$$

(This neglects losses, armature resistance and saturation.)

MACHINERY 5

A 100-kva, 11,000/2200-v., 60-cycle, single-phase transformer has an Hysteresis loss of 750 watts, an Eddy-current loss of 225 watts, and a copper loss of 940 watts under the rated conditions of full load. It is desired to export this transformer and to operate it at 45 cps but with the same maximum flux density and the same total loss as at 60 cps.

Calculate the new voltage and kva-rating. Neglect the exciting current.

SOLUTION

Use the expressions for Hysteresis and Eddy-current losses to obtain new values at the new frequency. The new ratings are determined from the values at the new frequency.

The total losses at full load and 60 cps are

$$P_T = P_h + P_e + P_{copper}$$

$$= 750 + 225 + 940 = 1915 \text{ watts}$$

The expressions for core loss are

$$P_h = K_h f B^n_{max}$$

$$P_e = K_e f^2 B^2_{max}$$

Let the primed values indicate values at the new frequency. The modified core losses are

$$P_h' = \frac{45}{60}(750) = 562 \text{ watts}$$

$$P_e' = \left(\frac{45}{60}\right)^2 (225) = 126.5 \text{ watts}$$

Assume the copper loss to be independent of frequency and flux density. Then.

$$P'_{copper} = P_T - P'_h - P'_e$$

$$= 1915 - 562 - 126.5 = 1226.5 \text{ watts}$$

If the voltage is sinusoidal

$$E = 4.44 B_{max} NA \times 10^{-8}$$

then

$$E'_p = \frac{45}{60}(11{,}000) = 8250\text{v.}$$

The rated primary current at 60 cps is

$$I_p = \frac{kva}{E} = \frac{100{,}000}{11{,}000} = 9.1 \text{ amp}$$

The total equivalent winding resistance referred to the high-voltage is 60 cps is

$$R = \frac{P_{copper}}{I^2_p} = \frac{940}{(9.1)^2} = 11.35 \text{ ohms}$$

The primary current at 45 cps is

$$I' = \sqrt{\frac{P'\ copper}{R}} = \sqrt{\frac{1226.5}{11.35}} = 10.4 \text{ amp}$$

The new kva rating is

$$kva' \quad \frac{(8250)(10.4)}{1000} = 86 \text{ kva} \qquad \text{Ans.}$$

The new voltage rating is

8250/1650 Ans.

NOTE: Alternate method to find I'_p.

$$\frac{P_{copper}}{P'_{copper}} = \frac{(9.1)^2 \not{R}}{I'^2_p \not{R}} = \frac{940}{1{,}226.5}$$

$$I_p' = \sqrt{\frac{1{,}226.5}{940}} \times 9.1 = 1.14 \times 9.1 = 10.4 \text{ amp}$$

MACHINERY 6

Two single phase motors are connected in parallel across a 120 volt, 60 cycle source of supply. Motor "A" is a split-phase induction type and motor "B" is a capacitor type.

Given the following data:

Motor	Horsepower Output	Motor Efficiency	Motor Power Factor
"A"	1/4	0.6	0.7 Lagging
"B"	1/2	0.7	0.95 Leading

REQUIRED:

Wt.

5 (a) Find the total power drawn, the combined line current and power factor of the two motors operating in parallel.

5 (b) Draw a vector diagram showing E_{line}, I_{line}, P_A, P_B, P_{total}. Label all items.

SOLUTION

a)

motor	HP_{op}	Eff.	Pf
A	1/4	0.6	0.7 lag
B	1/2	0.7	0.95 lead

Output Motor A = VI Cos Θ × Eff.

In this case

$$HP_{op} = \text{Fraction of load} \times \frac{746 \text{ watts}}{HP} = 0.25\,HP \times \frac{746}{HP} = 186.5 \text{ watts}$$

$$I_{LINE} = \frac{VI \cos\Theta}{V \times Eff. \times Pf} = \frac{0.25 \times 746}{120 \times 0.7 \times 0.6} = 3.7 \text{ amp.}$$

$$\text{Input Motor A} = \frac{\text{Output}}{Eff.} = \frac{186.5}{0.6} = 311 \text{ watts}$$

$$E_{LINE} = 120 \text{ volts}$$

Output Motor B = VI Cos Θ × Eff.

$$= 0.5\,HP \times \frac{746}{HP} = 373 \text{ watts}$$

$$I_{LINE} = \frac{0.5 \times 746}{120 \times 0.95 \times 0.7} = 4.66 \text{ amp.}$$

$$\text{Input Motor B} = \frac{\text{Output}}{Eff.} = \frac{373}{0.7} = 534 \text{ watts}$$

Motor	Pf	VA	Θ	Sin Θ	Input Power
A	0.7 lag	445	45°	0.715	311 watts
B	0.95 lead	562	18.2°	0.3123	534 watts

Total power = Input power A + Input power B

= 311 + 534 = 845 watts

Motor A VARS = 318 lag

Motor B VARS = −176 lead

Total VARS = 142 lag

$$\Theta' = Tan^{-1}\frac{142}{845} = 9.6^\circ$$

$$Cos\ \Theta' = cos\ (9.6^\circ) = 0.986 \text{ Line Power factor}$$

b)

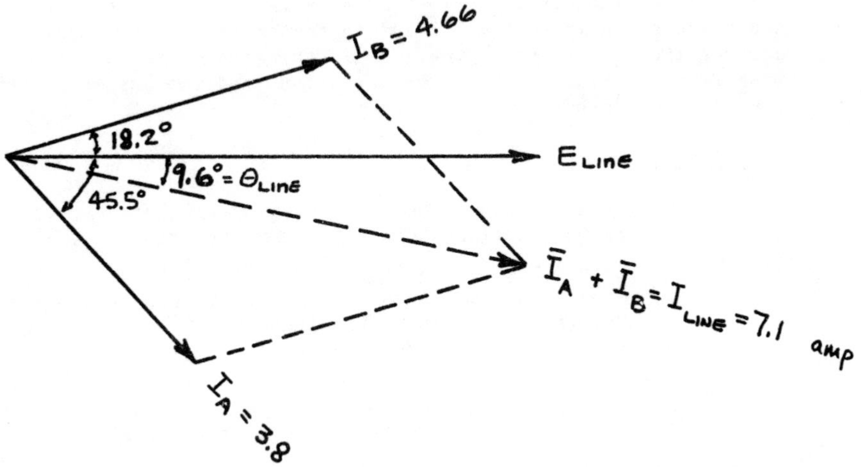

$$\bar{I}_{LINE} = I_A\,(Cos\,45.5^\circ - j\,Sin\,45.5^\circ) + I_B\,(Cos\,18.2^\circ + j\,Sin\,18.2^\circ)$$

$$= 3.7\,(0.7 - j\,0.71) + 4.66\,(0.95 + j\,0.3123)$$

$$= 2.59 - j\,2.62 + 4.43 + j\,1.42 = 7.02 - j\,1.2$$

$$I_{LINE} = 7.1 \text{ amp.}$$

$$P_{TOTAL} = P_A + P_B = 845 \text{ watts}$$

MACHINERY 7

In an emergency, a d.c. motor must be used as a generator.

The motor is a cumulative-compound motor; its efficiency is 90%, and its positive and negative terminals are marked.

The cumulative-compound characteristic must be maintained when it is used as a generator, the rotation must be kept in the same direction and the rpm will be the same. The positive terminal must remain the positive terminal in the operation as a generator. The interpoles must aid commutation in both motor and generator mode. The machine is to deliver the same power to the line in the generator-operation as it took from the line in the motor-operation, and the losses in the machine are the same in both cases.

REQUIRED:

(a) Determine if any of the following changes are necessary, and do enough calculations to verify your answer:

1. Should the shunt-winding connections be changed?
2. Should the compound-winding connections be changed?
3. Should the interpole-winding connections be changed?
4. Should the field resistance be changed?

(b) Has the load on the mechanical clutch of the machine increased or decreased when the operation changed from motor to generator if it was running at full capacity as a motor?

(c) How much does the electrical power transferred at the line connection increase or decrease if the same mechanical torque is maintained on the shaft?

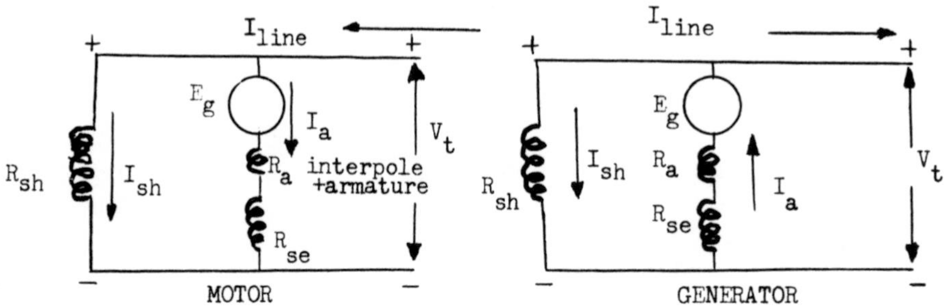

SOLUTION

(a) 1. No, the shunt winding should not be changed. The problem clearly states that direction of rotation and polarity (positive terminal) stay the same. Therefore regardless of whether we have a motor or generator operation, the direction of the current stays unchanged. The only drawback is the subtractive MMF of shunt and series field winding, giving a differential compounding. This matter is brought up in the next point.

2. Yes, the compound (series) winding connections should be changed, since the armature current during generator operations is reversed and now opposes (differentially compounded) the shunt field. To still aid (cumulatively compounded) the shunt field, the compound (series) winding should be reversed.
3. No, the interpole winding connections should not be changed. Interpoles or commutating poles are narrow laminated auxiliary poles placed midway between the main poles and the plane of commutation. These interpoles are in series with the armature and are wound to oppose and nullify the armature reaction in the commutating plane. This prevents sparking that might cause flashover and also reduces iron losses in the armature teeth.
   Changing from motor to generator action, the polarity of the commutating pole automatically changes, with the change of the armature reaction MMF. Therefore, commutation in interpole machines is not affected by a change from motor to generator operation or a change in the direction of rotation.
4. Yes, the field (shunt) resistance should be changed. Since the generator voltage has to be larger than the terminal voltage due to the ohmic voltage drop in the armature winding, the flux has to be increased ($E_G = K \cdot \varnothing \cdot$ rpm). Increased $\varnothing$ means increased field current. Therefore, the shunt field resistance should be decreased.

(b) The load on the mechanical clutch of the machine has increased.

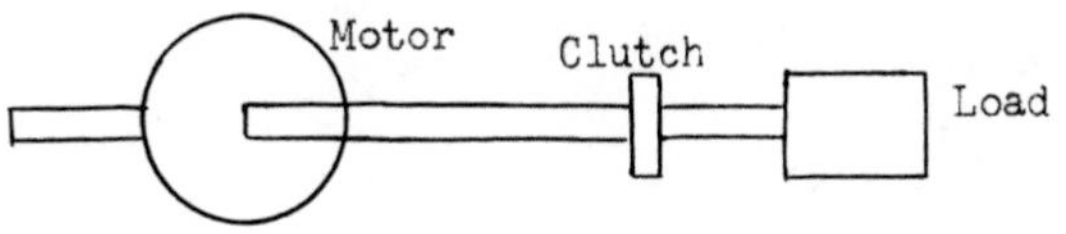

Rated Power_{input} = Losses + Load on clutch (for motor)

Rated Power_{output} + Losses = Load on clutch (for generator)

Since losses remain the same in either mode of operation, the load on the clutch increases when machine is operated as a generator.

(c) Using the line connections as reference point, due to internal losses, the power delivered to the motor shaft is 90% of the reference or input power.
Assuming that the same mechanical torque is maintained, above shaft power is now the input (prime mover) of the generator operation. Therefore, using the same 90% efficiency of the machine, the generator output referred to the original electrical power transferred at the line connections has decreased to: 0.90 x 0.90 = 0.81 or 81%

Reference: Electrical Circuits and Machines, Chapter XIII, by Robertson and Black.

MACHINERY 8

Two Y connected, 50° rise induction motors are fed by a 4160 V, line-to-line, 3-phase 60 Hz motor-control center 20 feet away. Motor #1 drives a 600 HP compressor. The efficiency of the motor is 90%, and its power factor is 0.5. Instruments of motor #2 indicate 1730 kw, 277 amps.

REQUIRED:

(a) Show the phasor-diagram of the loads, kw and kva.
(b) Determine the capacity in microfarads of a wye-connected capacitor bank that is required to correct the power factor of the total load to 0.966.
(c) If a synchronous motor is installed in place of motor #2 and used instead of the capacitor bank to achieve the same over-all power factor (0.966), what must its power factor be?
(d) Determine the feeder size (copper) and the rating of the control center circuit breaker, or fuse, that must be used for each of the following conditions, and indicate what section of the Electrical Safety Orders is applicable: (1) The 2 induction motors. (2) The 2 motors with the capacitor bank connected. (3) 1 induction motor and 1 synchronous motor.

SOLUTION

(a) Motor #1 = $\frac{600}{0.90} \times 0.746 = 497 \text{ KW} \cong 500 \text{ KW}$

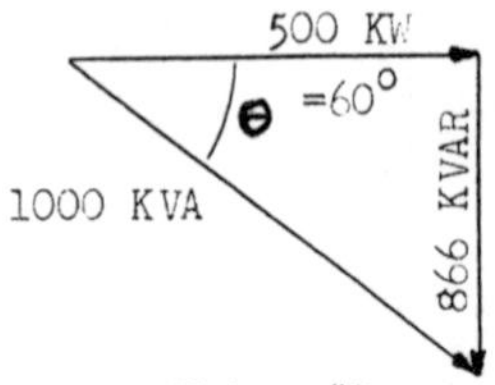

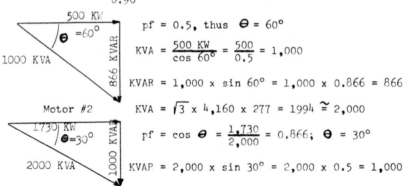

pf = 0.5, thus $\theta = 60°$

$\text{KVA} = \frac{500 \text{ KW}}{\cos 60°} = \frac{500}{0.5} = 1{,}000$

$\text{KVAR} = 1{,}000 \times \sin 60° = 1{,}000 \times 0.866 = 866$

Motor #2

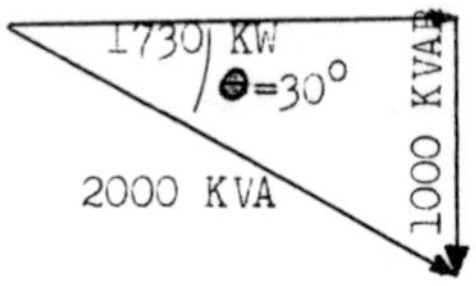

$\text{KVA} = \sqrt{3} \times 4{,}160 \times 277 = 1994 \cong 2{,}000$

$\text{pf} = \cos\theta = \frac{1{,}730}{2{,}000} = 0.866; \ \theta = 30°$

$\text{KVAR} = 2{,}000 \times \sin 30° = 2{,}000 \times 0.5 = 1{,}000$

(b) Total load of motors #1 and #2

$\text{KW} = 500 + 1{,}730 = 2{,}230$
$\text{KVAR} = 866 + 1{,}000 = 1{,}866$
$\text{KVA} = \sqrt{2{,}230^2 + 1{,}866^2} = \sqrt{8.45 \times 10^6} = 2{,}907$

Actual combined pf $= \cos\theta = \frac{2{,}230}{2{,}907} = 0.767$ lag; $\theta = 40°$

Desired combined pf = 0.966 lag; $\theta = 15°$

$$\text{KVA new} = \frac{2{,}230}{0.966} = 2{,}308$$

BC = Required leading KVAR = 1,866 - 2,230 tan 15°
  = 1,866 - 2,230 x 0.268 = 1,866 - 598 = 1,268

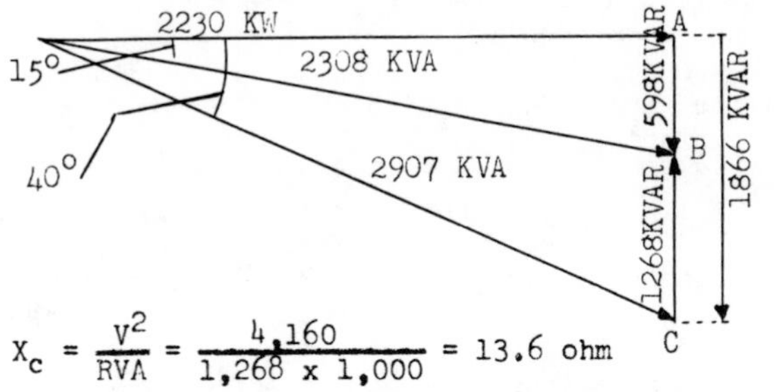

$$X_c = \frac{V^2}{RVA} = \frac{4{,}160}{1{,}268 \times 1{,}000} = 13.6 \text{ ohm}$$

$$C = \frac{1}{2\pi f \cdot X_c} = \frac{1}{6.28 \times 60 \times 13.6} = 195 \mu F$$ ANS.

(c) Assuming that synchronous motor has same efficiency as motor #2 which it replaces,

KVAR syn. motor = KVAR motor #1 - KVAR desired = AC - AB
= 866 - 598 = 268

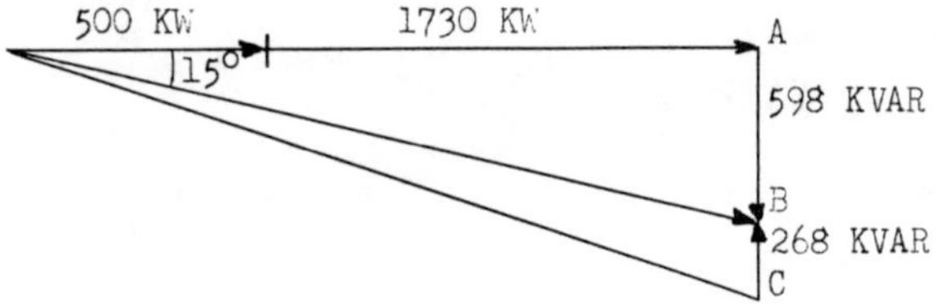

The synchronous motor alone:

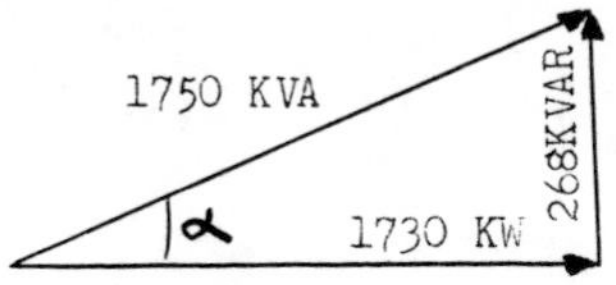

$$\tan \alpha = \frac{268}{1{,}730} = 0.155; \ \alpha = 9^\circ$$

$$\text{KVA} = \frac{1{,}730}{\cos} = \frac{1{,}730}{0.988} = 1{,}750$$

pf = cos 9° = 0.988 ANS.

(d)
(1) Two induction motors
A. For feeder size, from ESO 2395 (a), p. 261

$$1.25 \times I_{\text{motor \#2}} + I_{\text{motor \#1}} = 1.25 \frac{2{,}000{,}000}{\sqrt{3} \times 4{,}160} + \frac{1{,}000{,}000}{\sqrt{3} \times 4{,}160}$$

$$= 1.25 \times 279 + 139 = 348 + 139 = 487 \text{ amps}$$

From ESO Article 38 Table 1A p. 358.22 for non-continuous loads (i.e., for other than air conditioning motors) we select a 750 MCM feeder. ANS.

B. For overcurrent protection of motors from ESO 2396 (a), p. 262 and Article 38, Table 10, p. 358.22, column 8 (we assume squirrel cage motor without code letters, since none are given, and more than 30 amperes):

$I_{motor\ \#2}$ = 278 (the largest current) has breaker or fuse rating of 600 amps.

Total rating = 600 + $I_{motor\ \#1}$ = 600 + 139 = 739 amps

Since this is not a standard size, we use the next higher value, or 800 amps. ANS.

(2) The 2 motors with the capacitor bank.

A. $I_{total} = \frac{2,308,000}{\sqrt{3} \times 4,160} = 321$ amps

We assume that the capacitor is located at the motor.

$I_{motor\ \#1} = 321 \times \frac{500}{2,230} = 72$

$I_{motor\ \#2} = 321 - 72 = 249$

To select feeder:
1.25 x 249 + 72 = 311 + 72 = 383 amps

From Table 1A we select size 500 MCM or 2 sets of 250 MCM feeders. ANS.

B. $I_{motor\ \#2}$ = 249 ......500 amp breaker

$I_{motor\ \#1}$ = 72 amps

Total rating = 500 + 72 = 572 amps. Select nearest size of 600 amps. ANS.

(3)
A. $I_{syn\ motor} = \frac{1,750,000}{\sqrt{3} \times 4,160} = 243$ amps

$I_{motor\ \#1}$ = 72 amps

These values are very close to above case (2)A. We select same size 500 MCM or 2 sets of 250 MCM feeders. ANS.

B. Breaker rating is 600 amps; see above case (2)B. ANS.

5

Power Distribution

The few problems involving power distribution are somewhat more specialized. If one does not have a good feel for the following items the study time may better be spent on circuits, machines, controls, etc. A good understanding is needed in:

1. Fields
2. Transformer equivalent circuits
3. Per unit values
4. Synchronous machines

Because of the broader aspects of these types of problems only a few of the more often occurring relationships will be reviewed here.

For both this area and the machinery area, many of the problems are presented using per unit values. Consider for example, a name plate rating of a three phase device. Then, on a per phase basis, one could establish this value as the base rating and all other values could be considered to be a percentage of this base value. And, of course, a base voltage, current, or impedance is established the same way; thus when working with transmission lines feeding transformer banks, one can either work on the high or low voltage side as the per unit values allows for this transformation.

Fortunately, most of the power distribution problems do not involve transient analysis and many can be worked with a basic understanding of transformer connections and their equivalent circuits. One is reminded that a basic three phase system is usually broken down to a per phase relationship assuming wye connections. If the problem is one of the line having a particular impedance and feeding a bank of transformers, then set up the problem with the transformer equivalent circuit parameters referred to the transmission side of the transformer. Then, the overall effective impedance can usually be lumped together (on a per phase basis) to solve for the quantity sought.

If the problem is one of calculating transmission line impedances, fields, etc., then unless one has background in this area (including the use of the Smith chart), the particular problem might better be skipped. Generally, the transmission lines are characterized by circuit parameters both in shunt and in series. Of course the line resistance and the self-inductance are part of the series representation while the capacitive effect belong in the parallel portion of the representation (usually the shunt resistance may be neglected). This shunt, or capacitive effect, will be equal for all three phases (for phase symmetrical lines) and consequently the wye equivalent can be thought of as the neutral node being ground. Then for many of these problems, the line can be thought of as a lumped parameter equivalent circuit as follows:

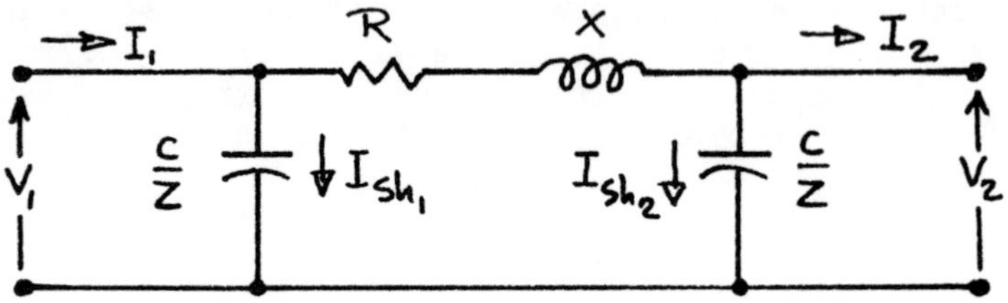

(For details of a solution for this type of problem, see page 262 of *Basic Electric Power Engineering* by Elgerd.)

In general, one can say that if a generator is placed on a large interconnected system, and the generator is made to increase its real power output, the system voltage change is greatest close to the generator while the effect diminishes at greater distances away from the generator (along the power grid).

POWER 1

A three-phase 173 KV, 50 mile long transmission power line has been tested for it's characteristic parameters, with the following results (on a per phase basis):

Resistance = 0.1 ohm/mile

Inductance = 2.0 millihenry/mile

Capacitance = $1.0 \times 10^{-2} \mu f$ /mile

Assume the received complex power at the end of the line is

$75 + j30$ (three-phase, total megavolt-amperes)

Find the input power.

SOLUTION

See Elgerd: *Basic Electric Power Engineering*. Addison-Wesley, 1977.

The equivalent circuit may be represented as:

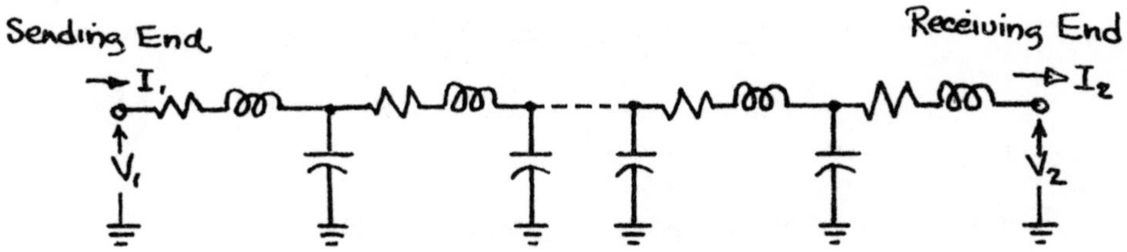

And a simpler approximation (on a per phase basis) is:

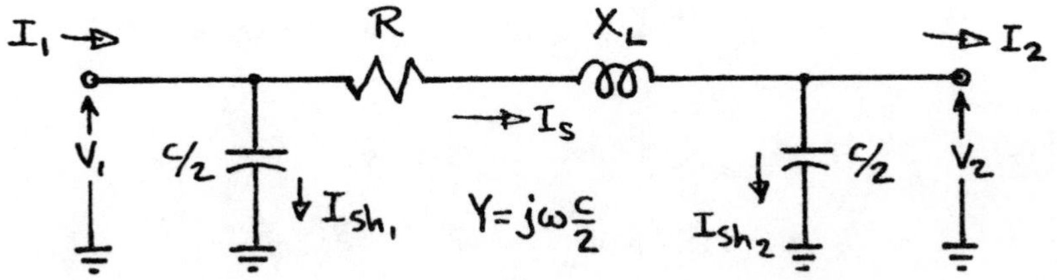

The complex power (which may also be expressed as the phasor voltage times the complex conjugate of the phasor current) on a per phase basis is $\frac{1}{3}(75 + j30) = 25 + j10$ mega volt-amps.

The current at the receiving end, then, is:

$$I_2 = \frac{(25+j10)\times 10^6}{(173/\sqrt{3})\times 10^3} = (0.25+j0.1)\times 10^3 \text{ A/phase}$$

The shunt and series currents are then found

$$I_{sh_2} = V_2 Y_2 \quad \text{where } Y_2 = j\tfrac{1}{2}(50)(377)(1\times 10^{-8})$$

$$= j94.25\times 10^{-6}$$

$$= (173/\sqrt{3})\times 10^3 (j94.25)\times 10^{-6}$$

$$= j9.425 \text{ A/Phase}$$

$$I_{ser} = I_{sh_2} + I_2 = j9.425 + 250 + j100$$

$$= 250 + j109.4 \text{ A/Phase}$$

And the voltage drop across the series branch may be found by first finding the series impedance:

$$Z = R + jX_L = (50)(0.1) + j(50)(377)(2\times 10^{-3})$$

$$= 5 + j37.7 \text{ ohms/phase}$$

$$V_{drop} = I_{ser} Z = (250+j109.4)(5+j37.7)$$

$$= -2.874 + j10.07 \text{ KV/phase}$$

Therefore the input voltage is

$$V_1 = V_{drop} + V_2 = -2.874 + j10.07 + 173/\sqrt{3}$$

$$= 97.13 + j10.07 \text{ KV}$$

And the sending end shunt current is

$$I_{sh_1} = V_1 Y = (97.13 + j10.07)(j94.25\times 10^{-6})$$

$$= 0.949 + j9.1545 \text{ A/phase}$$

Therefore the sending end current is

$$I_1 = I_{sh_1} + I_{ser} = 0.949 + j9.154 + 250 + j109.4$$
$$= 250.9 + j118.55 \text{ A/phase}$$

And then one may find the sending end complex power (again complex voltage times the complex conjugate of the current) as

$$\text{Complex power} = (97.13 + j10.07) \times 10^3 (250.9 - j118.55)$$
$$= 25.56 + j1.38 \text{ mega volt-amps/phase}$$

And, of course, the total complex power is three times this value :

$$76.68 + j4.14 \text{ mega volt-amps}$$

POWER 2

A 345 KV power transmission line has two bundled conductors per phase, spaced 18 inches horizontally. The conductor used in the bundle has a self GMD of 0.0403 feet and the phases are spaced horizontally 15½ feet apart.

REQUIRED:

Determine the following:
(a) The self GMD of the bundled conductors.
(b) The mutual GMD of the line.
(c) The inductive reactance per phase per mile.

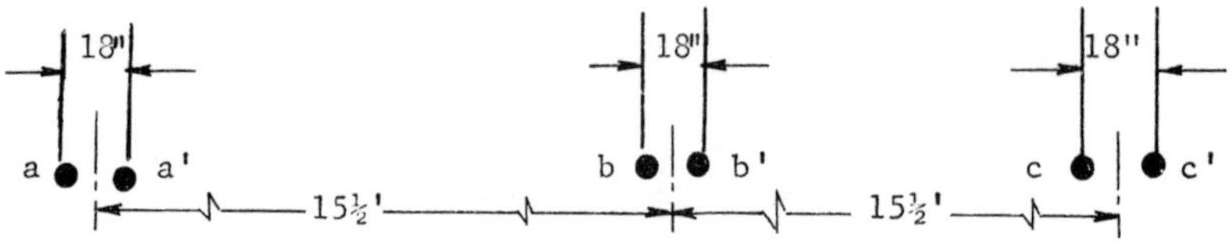

SOLUTION

(a)
Often self GMD of a bundled or composite conductor is called "geometric mean radius", or GMR. Self GMD may be denoted as D_s. This term includes the distances of a strand or conductor from all other strands within the same bundle plus the "distance of the strand from himself" or the self GMR of the strand.

In above line configuration we have two strands per bundle, thus we have four distances: $D_{aa'}$, $D_{a'a}$, D_{aa}, $D_{a'a'}$. The self GMR of a single strand is less than the actual physical radius (R x 0.7788). This reduced radius is the above given self GMD of 0.0403 ft. and is available from tables. Converting all distances into feet, we obtain for one bundle:

$$D_s = D_{sa} = \sqrt[4]{D_{aa'} \times D_{a'a} \times D_{aa} \times D_{a'a'}} = D_{sb} = D_{sc}$$

Note: We extract the fourth root as we have four distances under the radical

$$D_s = \sqrt[4]{(0.0403)^2 \times \frac{18}{12}^2} = \sqrt[4]{0.00366} = \frac{\log 0.00366}{4} =$$

$$= \frac{\log 3.66 - 4}{4} = \frac{0.564 - 3}{4} = \frac{1.564 - 4}{4} = 0.391\text{-}1$$

$$= 0.246 \text{ ft.} \quad \text{ANS.}$$

(b)
The mutual GMD of the line or D_{eq} is the geometric mean of all mutual GMD values outside the bundles, i.e. between the three phases.

$$D_{ab} = D_{bc} = \sqrt[4]{(15.5)^2 \times 17.0 \times 14.0} \text{ where ab} = 15.5'; \text{ a'b'} = 15.5'$$

$$ab' = 15.5 + 2 \times \frac{18}{12} \times \frac{1}{2} = 17.0'$$
$$a'b = 15.5 - 2 \times \frac{18}{12} \times \frac{1}{2} = 14.0'$$

$$= \sqrt[4]{57{,}180} = \frac{\log 57{,}180}{4} = \frac{4.758}{4} = 1.19 = 15.5 \text{ ft.}$$

$$D_{ac} = \sqrt[4]{(15.5 \times 2)^2 \times 32.5 \times 29.5} = \sqrt[4]{921{,}400} = \frac{\log\ 921{,}400}{4}$$

$$= \frac{5.966}{4} = 1.492 = 31 \text{ ft.}$$

$$D_{eq} = \sqrt[3]{D_{ab} \times D_{bc} \times D_{ac}} = \sqrt[3]{15.5 \times 15.5 \times 31.0} = 19.5 \text{ ft.} \quad \text{ANS.}$$

(c)
The inductive reactance in ohm/mile or $X_L = 2\pi f \times 10^{-3} \times 0.7411 \log \frac{Deq}{D_s}$. The 10^{-3} factor is needed to convert the inductance L, obtained in mh/mile into h/mile to finally yield ohm/mile.

$$X_L = 0.377 \times 0.7411 \log \frac{19.5}{0.246} = 0.279 \times \log 79.3 = 0.279 \times 1.90 =$$

$$= 0.530 \text{ ohm/mile} \quad \text{ANS.}$$

Reference: William D. Stevenson, Jr. "Elements of Power System Analysis", McGraw Hill. p. 30-38.

POWER 3

Power for a remote building on an industrial site is supplied through an existing buried cable from a fixed voltage 60-cycle supply.

The load in the remote building consists of lighting and induction motors. During periods of peak demand, when the cable is carrying approximately its rated current, the resulting steady-state load voltage is well below the desired value because of the characteristics of the load. A small amount of additional constant-speed motor load is anticipated in the near future.

REQUIRED:

What equipment can be installed at the building to improve the present situation and to permit the additional load? Explain how the equipment you recommend will improve the situation; the use of phasor diagrams is suggested.

SOLUTION

The crux of the whole problem is the large portion of induction motors. The power factor of induction motors at rated load are typically from 0.70 to 0.90 with some groupings of motors resulting in even lower power factors.

For a power factor of 0.8, a motor drawing 225 KVA of power will utilize only 180 KW

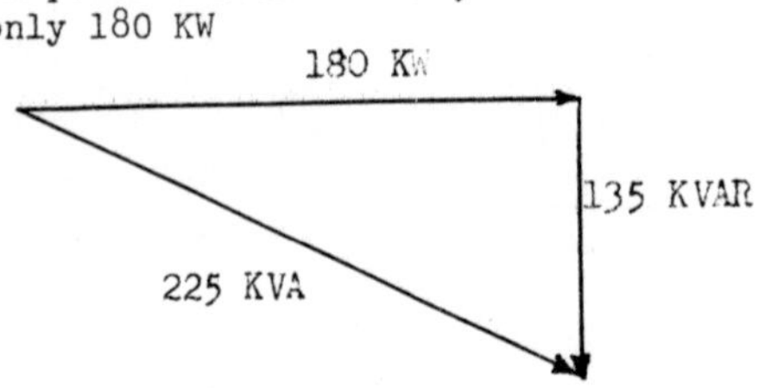

KW = Real power

KVAR = Reactive power

KVA = Apparent power

If the KVA drawn in this case were equal to the real power required (KW) a 20% reduction in current would result. The reduction in current with present load would reduce the voltage drop, thus improve the voltage at the load.

Fluorescent lighting with capacitors usually has power factors from 0.95 to 0.97, therefore are not practical to try to improve the power factor any higher.

To improve the power factor with the existing loads, capacitors should be applied. They have the characteristic of a leading power factor whereas induction motors have a lagging power factor. By adding capacitors their leading KVAR cancels out the equivalent amount of lagging KVAR, i.e.:

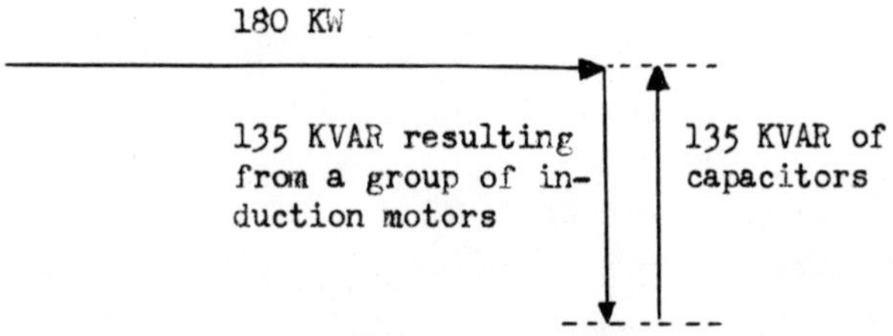

Above power factor correction results in 180 KVA = 180 KW or power factor = 1.0.

Another method of improving power factor is to add synchronous motors for the additional motor requirements. The synchronous motor acts like a capacitor producing a leading power factor and leading KVAR.

Normally capacitors are the most effective in reducing system costs when located near the devices with low power factor. Here we are primarily concerned with the feeder to the building, but if there are any large induction motors or a grouping of motors it would minimize local branch circuit voltage drop as well as the feeder, if the capacitors were located near the source of the low power factor (pf).

An economic study of the situation should be made. Data from recording pf meters and KW meters should be gathered from as many places as feasible on feeder and branch circuits. Then a comparison of how much and where the capacitors should be installed. The installation of the synchronous motors vs. induction motors with capacitors should be evaluated.

With the additional load the voltage drop in the existing feeder may be too much even with unity pf. Then consideration should be given to using boost transformer. It may be well that a combination of boosting and capacitor and synchronous motor will be the most economical solution. Boost transformers are much less in cost than regular transformers as they are just an auto-transformer. Improving the power factor beyond a certain point increases the cost disproportionately to the gain obtained, thus all alternates should be weighed in making the ultimate decision. Using a boost transformer alone may mean that at light load an overvoltage may result which could be undesirable.

Load profile diagram:

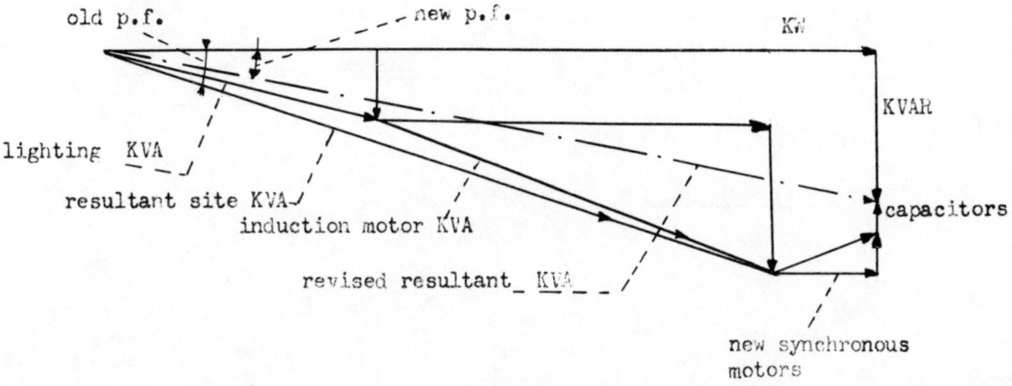

POWER 4

In the diagram below, determine the fault currents at point "F" for the following conditions:

Wt.

4 (a) A 3-phase fault.

6 (b) A single line to ground fault.

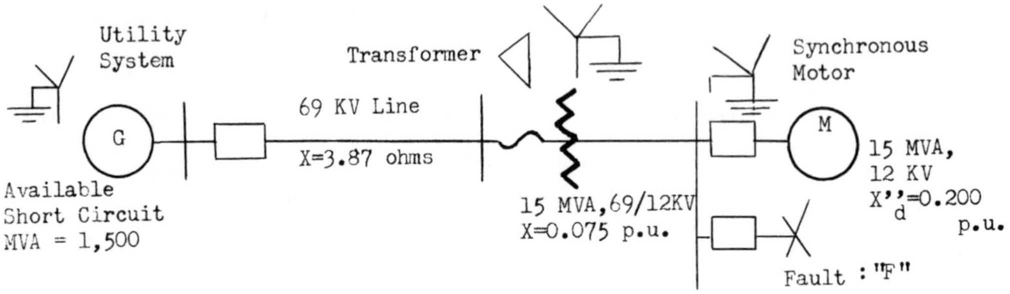

SOLUTION

KVA_{base} = 150 MVA (This value was selected to be a practical base between the two given MVA values)

KV_{base} = 69 and 12 respectively

$$Z_{base\ 69} = \frac{(KV)^2 \times 1000}{KVA_{base}} = \frac{(69)^2 \times 1000}{150{,}000} = 31.8 \text{ ohms}$$

$$Z_{base\ 12} = \frac{(12)^2 \times 1000}{150{,}000} = 0.96 \text{ ohms}$$

$$I_{base\ 69} = \frac{KVA_{base}}{\sqrt{3}\ KV_{base}} = \frac{150{,}000}{\sqrt{3} \times 69} = 1{,}250 \text{ amps}$$

$$I_{base\ 12} = \frac{150{,}000}{\sqrt{3} \times 12} = 7{,}230 \text{ amps}$$

$$Z_{line\ p.u.} = \frac{Z_{rated\ ohms}}{Z_{base}} = j\frac{3.87}{31.8} = j0.121 \text{ p.u.}$$

$$Z_{trans\ p.u.} = Z_{rated\ p.u.} \frac{KVA_{base}}{KVA_{rated}} = j0.075 \frac{150{,}000}{15{,}000} = j0.750 \text{ p.u.}$$

$$Z_{utility} = 1.0 \frac{KVA_{base}}{KVA_{sh.ckt.}} = j \times 1.0 \frac{150{,}000}{1{,}500{,}000} = j0.100 \text{ p.u.}$$

$$Z_{motor\ dl''} = j0.200 \frac{150{,}000}{15{,}000} = j2.00 \text{ p.u.}$$

(a) Three phase fault

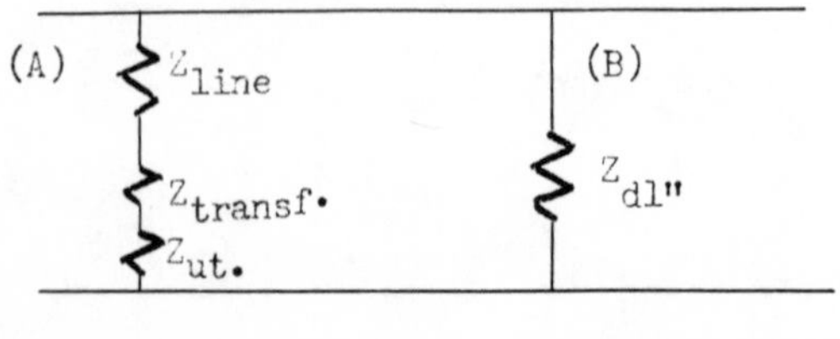

$Z_A = Z_{line} + Z_{transf} + Z_{ut} = j0.121 + j0.750 + j0.100 = j0.971$

$Z_B = Z_{d1''} = j0.200$

$$Z_{eq} = \frac{1}{j0.971} + \frac{1}{j0.200} = \frac{j0.971 \times j0.200}{j0.971 + j0.200} = j0.655$$

$$I_{fault} = \frac{E}{Z_{eq}} = \frac{1.0}{j0.655} = -j1.525 \text{ amps}$$

I_{fault} 3 ph at 12 KV $= I_{fault\ p.u.} \times I_{base}$ 12 KV

$= - j1.525 \times 7{,}230 = - j11{,}000$ amps ANS.

(b) Single phase fault

Positive sequence impedance diagram:

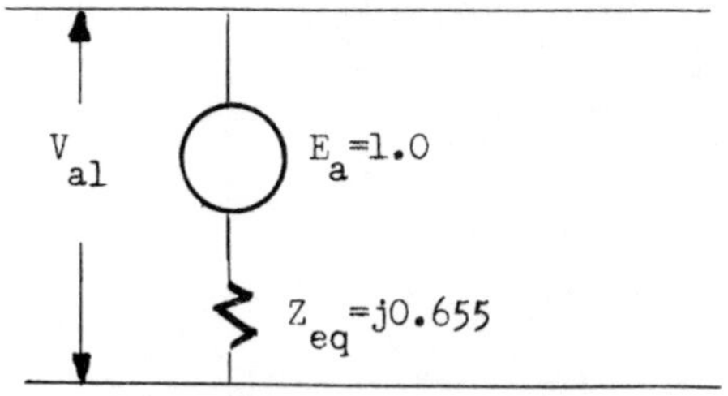

Negative sequence impedance diagram:

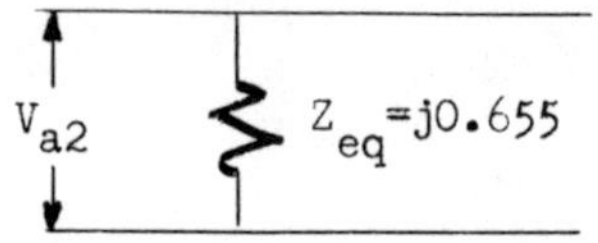

Zero sequence impedance diagram:

69 KV system zero sequence fault currents are isolated from fault "F" by △Y transformer

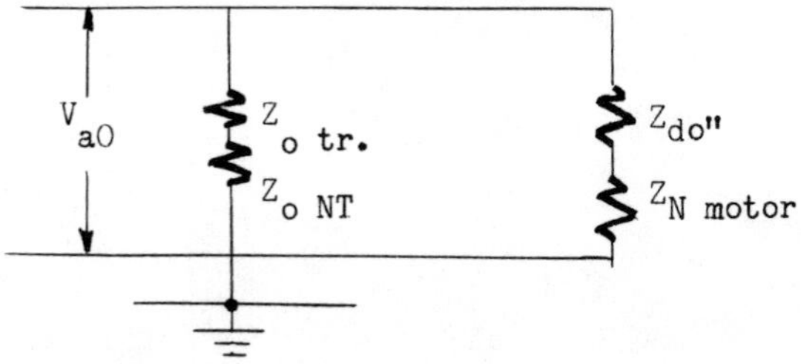

$Z_{o\ tr} = j0.750$

$Z_{motor\ do''} = \frac{1}{2} Z_{motor\ d1''}$ (assumed) $= j1.000$

$Z_{o\ NT} = 3Z_{N\ motor} = 0$ .... directly connected neutrals

$$Z_{o\ eq} = \frac{j0.750 \times j1.000}{j0.750 + j1.000} = j0.428$$

SEQUENCE NETWORK

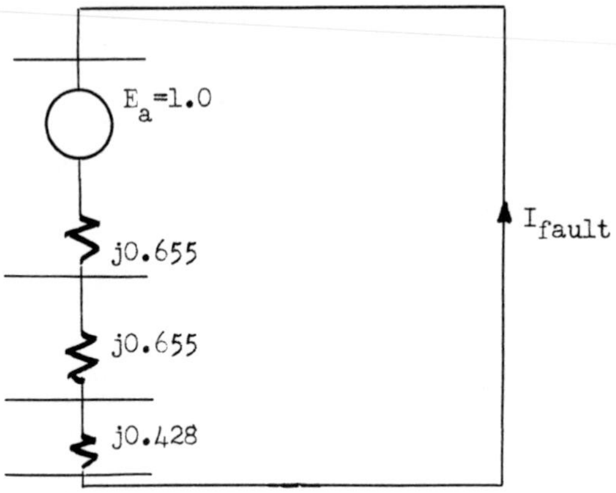

$$I_{a1} = I_{a2} = I_{ao} = \frac{E_a}{Z_1 + Z_2 + Z_o + 3Z_N + 3Z_{fault}}$$

Where $Z_N = 0$ and $Z_{fault} = 0$

thus:

$$I_{a1} = I_{a2} = I_{ao} = \frac{1.0}{j(0.655 + 0.655 + 0.428)} = -j0.580$$

$I_{fault\ p.u.} = (I_{a1} + I_{a2} + I_{ao}) = 3\ (-j0.580) = -j1.740$

$I_{fault} = I_{fault\ p.u.} \times I_{base\ at\ 12\ KV}$

$= -j1.740 \times 7{,}230 = -j12{,}500$ amps ANS.

POWER 5

An emergency 120/208-volt, 3-phase, 4-wire, 60-cps generator supplies an external circuit.

The load on the external circuit consists of 9000 watts of incandescent lights connected between line and neutral and evenly distributed among the 3 phases, and a 10-HP, 3-phase air conditioner motor of 83% efficiency and 0.707 P.F.

REQUIRED:

Wt.

5 (a) Show the phasor-diagram of the currents and voltages on the load-side of the generator.

3 (b) Determine the microfarads of the capacitor to be connected to the generator in order to reduce the generator load-current to 105% of that which would flow if the P.F. = 1.

2 (c) Determine the size of the conduit and wire to be used as a feeder and the required fuse size to protect the feeder when it is run between the generator and its distribution board in the next room, if the total load is continuous, and the capacitor calculated in (b) is connected.

SOLUTION

(a) Designating: I_m = Motor current; I_{mR} = Real component of I_m; I_{mQ} = Quadrature component of I_m; I_i = Incandescent lights current; I_1 = Total load current; I_1' = Corrected total load current; I_c = Capacitor current,

We obtain as follows:

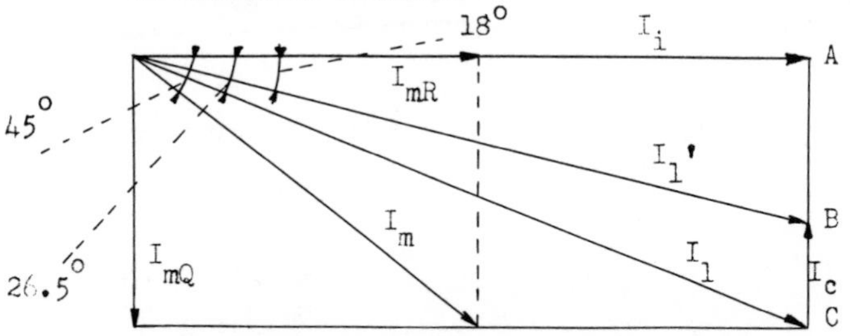

(b) $I_i = \dfrac{9{,}000}{\sqrt{3} \times 208} = 25\angle 0^\circ$ amps

$I_m = \dfrac{10 \times 746}{\sqrt{3} \times 208 \times 0.83 \times 0.707} = 35.3\angle 45^\circ$ amps = 25 + j25

Thus $I_{mR} = 25\angle 0^\circ$

$I_{mQ} = 25\angle -90^\circ$

$I_1 = 25 + 25 + j25 = 50 + j25 = 56\angle 26.5^\circ$ amps

If p.f. = 1, then I_1 reduces to the real component of above current i.e. $50\angle 0^\circ$. Therefore, we have to reduce $I_1 = 56\angle 26.5^\circ$ to 105% of 50 amps = $I_1' = 52.50$ amps at an angle to be determined.

From above vector diagram I_1' (52.50) ends in point B

$AC = 25$

$AB = \sqrt{52.5^2 - 50.0^2} = 16.5$

$BC = 25 - 16.5 = 8.5 \text{ amps} = I_c$

Then: $X_c/\text{phase} = \dfrac{E}{\sqrt{3} \times I_c} = \dfrac{208}{\sqrt{3} \times 8.5} = 14.1$ ohms

If $X_c = \dfrac{1}{2\pi fC}$

Then $C = \dfrac{1}{377 \times 14.1} = 18.7\ \mu F/\text{ph.}$

$C_{total} = 3 \times 18.7 = 56.1\ \mu F$ ANS.

(c) $I_1' = (25 + 25) + j16.5 = 52.50\angle 18^\circ$ amps

$I_{m\ new} = 25 + j16.5 = 30\angle 34^\circ$ amps

Rating of wire size:

$125\%\ I_{m\ new} + I_i = 1.25 \times 30 + 25 = 62.5$ amps

From copper wire tables (ESO*, Section 358.22 Table 1A) for 70 amps continuous current we obtain Size No. 4 ANS.

From over current protection table (ESO, Section 358.30 **Article** 38) at 30 amps (no 125% factor is needed) + 25 = 55 amps we obtain a fuse size of 120 amps ANS.

From conduit table (ESO, Section 358.26 Article 38, Table 4) for 4 wires and 55 amps we obtain a conduit size of 1 1/4" ANS.

*ESO refers to California's *"Electrical Safety Orders."* However, much of the material parallels the *"National Electrical Code,"* Underwriters Laboratories (National Board of Fire Underwriters, 85 John Street, New York, NY 10038).

POWER 6

There are presently available several different types of systems for use as 69 KV underground transmission lines. These systems are:

LPOF Single conductor cable

LPOF Three conductor cable

Solid type single conductor cable

MPGF

HPOF

HPGF

REQUIRED:

(a) Briefly describe each system, and list conditions that would favor the use of one system over another. Justify your answers.

(b) Which system is the most reliable?

(c) Which system is the least reliable?

(d) Which system costs the least (to install)?

(e) Which system probably costs the most for short runs?

SOLUTION

(a) 1. Low Pressure Oil-Filled Single Conductor Cable (LPOF).

Metallic oil channels are provided in place of filler material between the conductor and the load sheath. The oil-filled cable was developed to overcome the migration of compound and resultant voids inherent in solid cables by maintaining a static head of oil on the cable at all times. This is done by means of reservoirs. The mechanical strength of the lead sheath limits the static head permissible from 30 to 40 feet. Also, because of resistance to flow of oil, there are certain limitations on the length of a section of cable. Special stop joints are required between sections supplied by different reservoirs, and special terminating equipment must be used. When filling the oil system, special care must be exercised to thoroughly dry and de-gasify the oil.

2. Low Pressure Oil-Filled Three Conductor Cable (LPOF).

This system is similar to that of (a)1. above, with the exception that the cable contains 3 conductors instead of 1, and is less expensive. Three conductor oil-filled cable may be round or sector type and may be used on voltages from 20-69 KV between phases.

3. Solid Type Single Conductor Cable.

Impregnated paper insulation is wrapped around a single solid conductor cable. It cannot dissipate heat as well as oil- or gas-filled cables and therefore has a much shorter life. However, the cable is relatively inexpensive. Single conductor, solid type of paper - insulated cable may be used up to 70 KV, whereas the 3-conductor cables are rarely used above 35 KV because of excessive size and weight in the higher voltages.

4. <u>Medium Pressure Gas-Filled Cables (MPGF).</u>

   This cable is filled with inert gas (hydrogen) under a pressure of about 200 pounds per square inch. The inert gas fills the space between the paper insulted cable and the lead sheath. This cable has the advantage of being cooled without maintaining a reservoir. Further, the inert gas encounters very little resistance to flow.

5. <u>High Pressure Oil-Filled Cables (HPOF).</u>

   This system is similar to that of (a)2. above, with the advantage that oil under greater pressure has higher dielectric strength. However, the mechanical problems are increased. Special stop joints and terminating equipment are required to withstand the high pressures.

6. <u>High Pressure Gas-Filled Cables (HPGF).</u>

   This system is similar to that of (a)4. above, with the advantage that the inert gas under greater pressure has higher dielectric strength. However, the mechanical problems are increased.

(b) In general, the high pressure gas-filled cable system is the most reliable. It has comparatively few sources of mechanical problems.

(c) In general, the solid type single conductor system is the least reliable. Because of its poorer ability to dissipate heat, it is subject to higher voltage stresses, resulting in a shorter life.

(d) The solid type single conductor system would cost the least to install.

(e) In general, the high pressure gas-filled system would be the most expensive for short runs.

Reference: Standard Handbook for Electrical Engineers by A. E. Knowlton, Editor in Chief, Section 13, paras 210 to 220.

POWER 7

The voltages of an unbalanced 3-phase supply are $V_a = (200 + j0)V$, $V_b = (-j200)V$ and $V_c = (-100 + j200)V$.

Connected in star across this supply are three equal impedances of (20 + j10) ohms. There is no connection between the star point and the supply neutral.

REQUIRED:

Evaluate the symmetrical components of the A phase current and the three line currents.

SOLUTION

The voltage components are as follows:

$$V_1 = \frac{1}{3}(V_a + aV_b + a^2Vc)$$
$$= \frac{1}{3}\left[200 + (-0.5 + j0.866)(-j200) + (-0.5 - j0.866)(-100 + j200)\right]$$
$$= \frac{1}{3}(200 + j100 + 173.2 + 50 - j100 + j86.6 + 173.2)$$
$$= \frac{1}{3}(596.4 + j86.6) = 198.8 + j28.86$$
$$V_2 = \frac{1}{3}(V_a + a^2V_b + aV_c)$$
$$= \frac{1}{3}\left[200 + (-0.5 - j0.866)(-j200) + (-0.5 + j0.866)(-100 + j200)\right]$$
$$= \frac{1}{3}(200 + j100 - 173.2 + 50 - j86.6 - j100 - 173.2)$$
$$= \frac{1}{3}(-96.4 - j86.6) = -32.13 - j28.86$$
$$V_o = \frac{1}{3}(V_a + V_b + V_c) = \frac{1}{3}(200 + j0 - j200 - 100 + j200)$$
$$= \frac{1}{3}(100) = 33.33$$

$V_{a1} = 198.8 + j28.86 = (20 + j10)\, I_{a1}$

$V_{a2} = -32.13 - j28.86 = (20 + j10)\, I_{a2}$

$V_{ao} = 33.33 = \infty \cdot I_{ao}$, since neutral is not connected, i.e., there is no connection between the star point and the supply neutral.

$$I_{a1} = \frac{198.8 + j28.86}{20 + j10} \quad \text{ANS.}$$

$$I_{a2} = \frac{-32.13 - j28.86}{20 + j10} \quad \text{ANS.}$$

$$I_{ao} = \frac{33.33}{\infty} = 0 \quad \text{ANS.}$$

$$I_a = I_{a1} + I_{a2} + I_{ao} = \frac{1}{20 + j10}(198.8 + j28.86 - 32.13 - j28.86)$$

$$= \frac{1}{20 + j10}(166.67) = \frac{20 - j10}{500} \times 166.67 = I_a = \underline{\underline{6.67 - j3.33 \text{ amps in line a}}}$$ ANS.

To obtain the other (b,c) line currents:

$$I_b = a^2 I_{a1} + aI_{a2} + I_{ao}$$

$$= \left[(-0.5 - j0.866)(198.8 + j28.86) + (-0.5 + j0.866)(-32.13 - j28.86)\right]\frac{1}{20 + j10}$$

$$= (-99.40 - j172.16 - j14.43 + 24.99 + 16.65 - j27.82 + j14.43 + 24.99)\ \frac{20 - j10}{500}$$

$$= (-33 - j200)\ \frac{20 - j10}{500} = \underline{\underline{-5.33 - j7.33 \text{ amps in line b}}}$$ ANS.

$$I_c = a\ I_{a1} + a^2\ I_{a2} + I_{ao}$$

$$= \left[(-0.5 + j0.866)(198.8 + j28.86) + (-0.5 - j0.866)(-32.13 - j28.86)\right]\frac{1}{20 + j10}$$

$$= (-99.40 + 172.16 + j14.43 - 24.99 + 16.65 + j27.82 - j14.43 - 24.99)\ \frac{20 - j10}{500}$$

$$= (-133.33 + j200)\ \frac{20 - j10}{500}$$

$$= \underline{\underline{-1.33 + j10.67 \text{ amps in line c}}}$$ ANS.

As a check: $I_a + I_b + I_c = 0$, or

$6.67 - j3.33 - 5.33 - j7.33 - 1.33 + j10.67 = 0$

Reference: Elements of Power Systems Analysis by William D. Stevenson, Jr. Chapter 13.

6

Electronics

Problems in the electronics area primarily involve proper modeling of active devices such as transistors (both bi-polar and FET types). For the bi-polar types, the most frequently used parameter set to show up on typical exam problems is the hybrid or "*h*" parameter set.

DEFINITION:

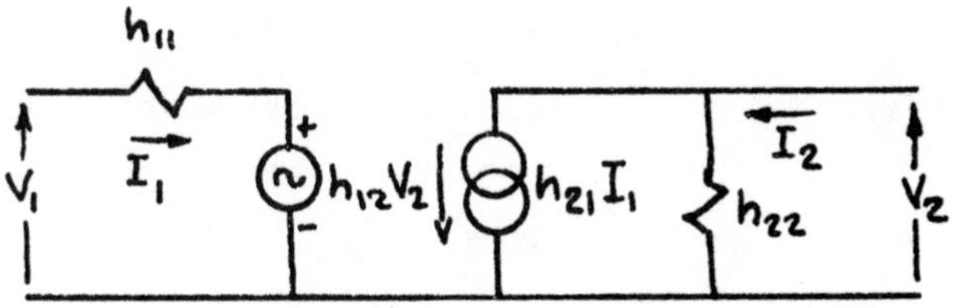

The defining equations are,

$$V_1 = h_{11} I_1 + h_{12} V_2$$

$$I_2 = h_{21} I_1 + h_{22} V_2$$

The h_{11} parameter is an input impedance value and is usually given as h_{ix} where the x can be e, c, or b depending on which terminal of the transistor is common to input and output. Thus $h_{ie} = h_{11}$, the input impedance value for a transistor in common emitter configuration.

h_{12} is a reverse voltage feedback ratio and thus h_{rb} is h_{12}, the reverse feedback ratio for a transistor in common base configuration.

h_{21} is a forward current transfer ratio, so h_{fc} would be the forward current transfer ratio for a common collector configuration. Finally, h_{22} is an output admittance value so h_{oe} is the output admittance for a common emitter transistor while h_{ob} is for a common base, etc.

The h-parameters are convenient to use because the gain or impedance equations are the same for all configurations. The drawback is that a different set of parameters must be used for each configuration. Ref: Linvill and Gibbons: *Transistors and Active Circuits,* McGraw-Hill, 1961. Pages 215-240. The h-parameters given are usually real quantities, although at higher frequencies they are complex.

A better model of the bipolar transistor for high frequencies or broad band amplifiers is the hybrid-Pi circuit. This circuit is developed in example problem 4. Its chief advantage is in broad band applications because it is simple to use and may be made unilateral.

For FET circuits, the most frequently used model to occur in these examinations, the open gate and transconductance current generator is in parallel with a drain resistance at the output. Biasing of the FET requires knowledge of the relationship of the drain saturation current, the pinch-off voltage, and the gate to source voltage. For an example see problem 3.

Simplifications to most electronics problems in these exams can be made by keeping in mind various impedance levels and neglecting terms which are large or small by comparison. In general, the output impedance of bi-polar transistors is reasonably high, in the order of 20K to 50K ohms for the common emitter configuration. This makes the h_{oe} term negligible when collector loads are an order of magnitude less. Similarly, the h_{re} term can usually be neglected when the collector loads are small. The elimination of the h_{oe} and h_{re} parameters reduces the low frequency model of the transistor to a very simple circuit allowing for a fast approximation of the overall gain and impedance values of an amplifier. These quick estimates are usually within 10% of the values obtained by using the more complete model and parameter values are seldom known to 10% accuracy.

In transistor switching circuits, a very good approximation to determine logic functions is to consider an on transistor as a three terminal short circuit (saturated condition) and an off transistor as three open circuited terminals. The saturated condition must be verified by checking the resistor values and power supply voltages against the minimum current gain specified for the transistors. If the I_c/I_b ratio is smaller than the minimum h_{fe} specified, then saturation can be assumed and the logic levels will be easily determined.

In conclusion, any linear electronics circuit can be analysed in stages by starting with the output stage and first determining the effective load seen by the stage. Then determine the Thevenin equivalent source impedance that drives the stage, and subsequently computing the stage gain. Then using a similar technique, work on the stage that drives the output stage and so forth.

If the amplifier has feedback, calculate the open loop gain as above with the feedback disconnected, but allowing for the loading effects of the feedback network. Then determine the feedback type, voltage or current, series or shunt, and the feedback ratio. You may then predict the closed loop behavior in accordance with the classical feedback relationship:

$$G = \frac{A}{1+AF}$$

where A is the open loop gain and F is the feedback ratio.

The effect of feedback on input and output impedances is similar to that of gain modification by the factor *(1 + AF)*. Care must be used to determine whether the factor increases or decreases the quantity in question. The following rules should help.

If the feedback is derived from the output voltage, then the output impedance is reduced by this factor. If the feedback is derived from the output current, then the output impedance is increased by this factor.

With respect to the input, if the feedback voltage is in series with the input source, the impedance is increased and if it is in shunt with the input, the input impedance is lowered. The classical example of the latter is the operational amplifier used in the inverting mode, the input terminal is a virtual short circuit to ground due to the large magnitude of the loop gain.

The problems in this section are typical of the problems found in many previous examinations and attempt to cover as many different types as possible. However, the broad field of electronics makes it impossible to provide examples of every type of problem that may occur.

ELECTRONICS 1

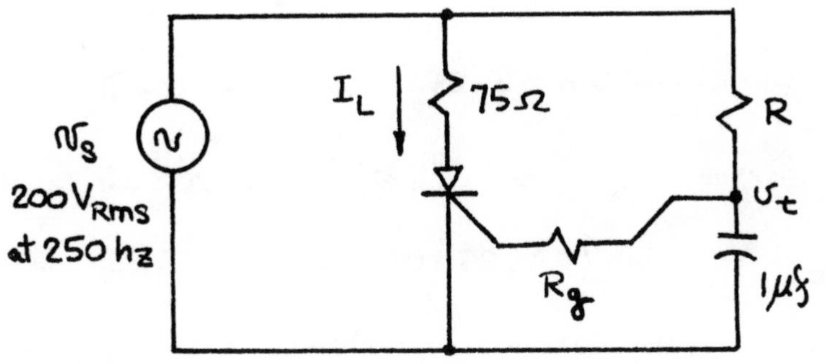

The SCR is to be considered ideal in the sense that it has a zero voltage drop while conducting and requires a negligible gate current to turn on. Determine a proper value of R so that I_L will be 1 amp dc. (average value current).

SOLUTION

First, the waveform of the load current must be determined so that the necessary phase angle for the conduction of the SCR can be determined. We sketch three pertinent waveforms and show their relative phases.

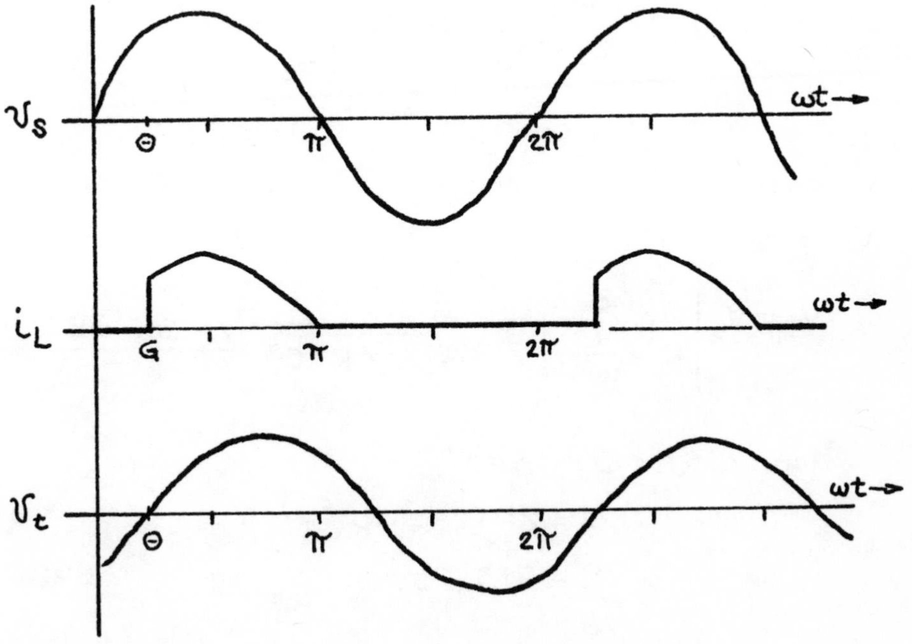

Since $V_{Rms} = \frac{V_{max}}{\sqrt{2}}$, $v_s = \sqrt{2}(200)\sin 2\pi(250)t$ volts

$$i_L = \frac{200\sqrt{2}}{75}\sin 2\pi(250)t, \qquad \theta < \omega t < \pi$$

$$i_L = 0 \qquad 0 < \omega t < \theta, \quad \pi < \omega t < 2\pi$$

$$v_t = V_{T_{max}}\sin(2\pi \times 250t - \theta)$$

v_t is the triggering voltage and goes positive at the point of the SCR firing $(\omega t = \theta)$. θ, the firing angle is determined from the requirement that $I_L = 1$ amp average.

$$I_{av} = \frac{1}{2\pi}\int_0^{2\pi} i(t)dt \quad \text{by definition}$$

$$\text{Then } 1 = \frac{1}{2\pi}\int_\theta^\pi \frac{200\sqrt{2}}{75}\sin(\omega t)\, d\omega t$$

$$1 = \frac{200\sqrt{2}}{2\pi \times 75}\left[-\cos \omega t\right]_\theta^\pi$$

$$1.67 = [1 + \cos\theta] \qquad \theta = 48.23°$$

Now determine the phasor value of V_T from the phasor value of $v_s \rightarrow V_s$

$$V_T = V_s\left[\frac{\frac{1}{j\omega C}}{R + \frac{1}{j\omega C}}\right] = \frac{V_s}{\sqrt{1+\omega^2R^2C^2}} \angle -\tan^{-1}\omega RC$$

$$\tan^{-1}\omega RC = \theta = 48.23°$$

$$\omega RC = 1.12 = 2\pi(250)10^{-6}R$$

$$R = 713\ \Omega$$

ELECTRONICS 2

TRANSISTOR AMPLIFIER PROBLEM USING h PARAMETERS

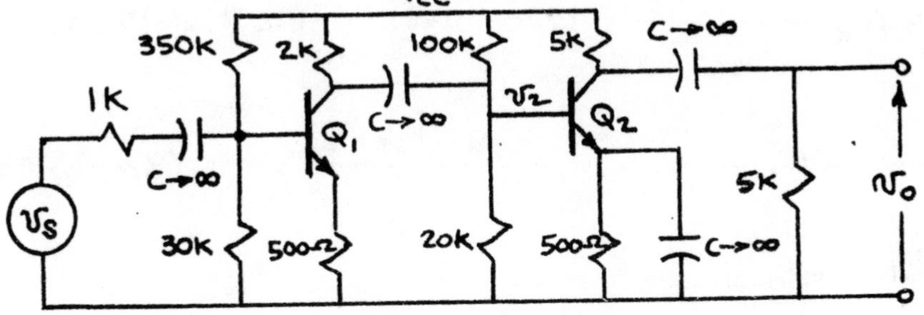

$h_{ie}\,(h_{11}) = 1K\,\Omega, \quad h_{fe}\,(h_{21}) = 50, \quad h_{re}\,(h_{12}) = 2\times10^{-4}$

$h_{oe}\,(h_{22}) = 20\times10^{-6}\,\mho$

Find v_o/v_s State any assumptions made.

SOLUTION

Using the h parameter model we will first find the voltage gain of stage 2, v_o/v_2 and the input impedance looking into Q_2. Then we can find the gain of stage 1 as loaded by stage 2.

For stage 2, the equivalent circuit in h parameter form:

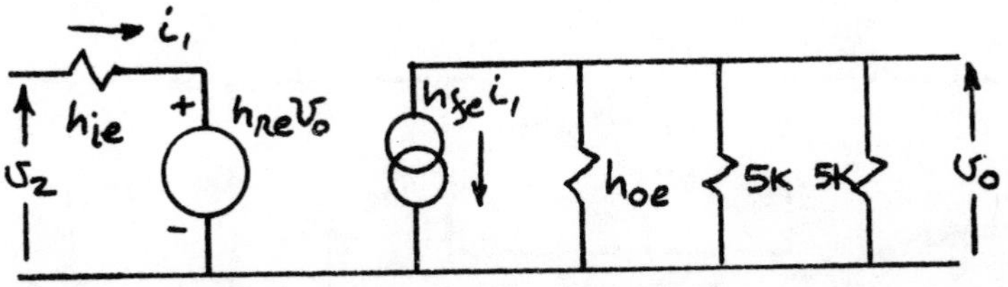

Let $Y_L = \dfrac{1}{5K /\!/ 5K}$ $\qquad Y_L = 400\times10^{-6}\,\mho$

I $\quad v_o = \dfrac{-h_{fe}\,i_1}{h_{oe} + Y_L} = -119\times10^{3}\,i_1$

II $\quad v_2 = h_{ie}\,i_1 + h_{re}\,v_o$

Then $v_2 = i_1 h_{ie} - \dfrac{i_1 h_{fe} h_{re}}{h_{22} + Y_L} = i_1 \left[h_{ie} - \dfrac{h_{fe} h_{re}}{h_{22} + Y_L} \right]$

$$Z_{in_2} = h_{ie} - \frac{h_{fe} h_{re}}{h_{22} + Y_L} = 1000 - 119 \times 10^3 (2 \times 10^{-4}) = 976\,\Omega$$

From I above, we have $i_1 = -\dfrac{v_o (h_{oe} + Y_L)}{h_{fe}} = -v_o (8.4 \times 10^{-6})$

Substituting into II gives

$$v_2 = h_{ie} \left[\frac{-v_o (h_{oe} + Y_L)}{h_{fe}} \right] - h_{re} v_o$$

$$= -(1K)(8.4 \times 10^{-6}) v_o - 2 \times 10^{-4} v_o$$

$$= -8.6 \times 10^{-3} v_o$$

$$\frac{v_o}{v_2} = -116$$

Now we take an equivalent circuit for the input stage including the loading by the input to the second stage and the biasing networks.

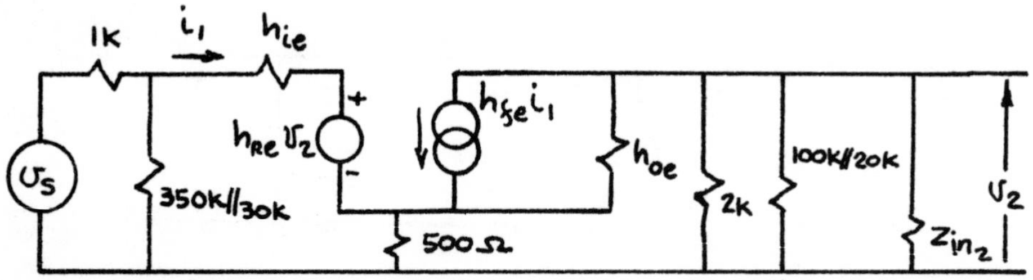

We note the fact that the 2K collector load is in parallel with Z_{in2} (976Ω) and these are in parallel with the bias network for transistor #2. Note that 100K‖20K = 16.7K while 2K‖976Ω = 656Ω. Therefore assume bias network for stage #2 can be neglected.

The unbypassed emitter resistor in stage #1 presents a very high input impedance to the source $(Z_{in} = h_{ie} + (h_{fe}+1)R_E)$. However, the h_{oe} of the transistor is high compared to the 500Ω so we may assume that we can neglect h_{oe} as being a high resistance [$1/h_{oe}$ = 50K] compared to either the load (656Ω) or or the 500Ω R_E. Finally, we will incorporate the bias network for transistor #1 into the source by taking a Thevenin equivalent CKT.

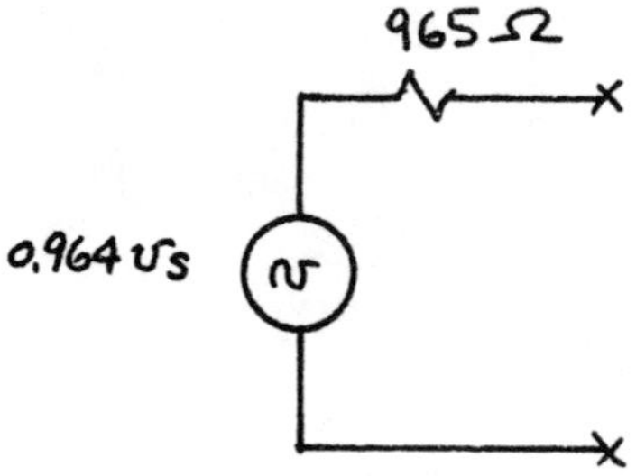

$$v_{Th} = v_S \frac{(350K//30K)}{(350K//30K)+1K} = 0.964\ v_S$$

$$R_{Th} = 1K // 30K // 350K = 965\ \Omega$$

By the previous discussion we will refer the R_E (500Ω) resistor to the input by $R_E' = (h_{fe} + 1)R_E = 25.5\text{K}\Omega$

Now our equivalent circuit looks like this:

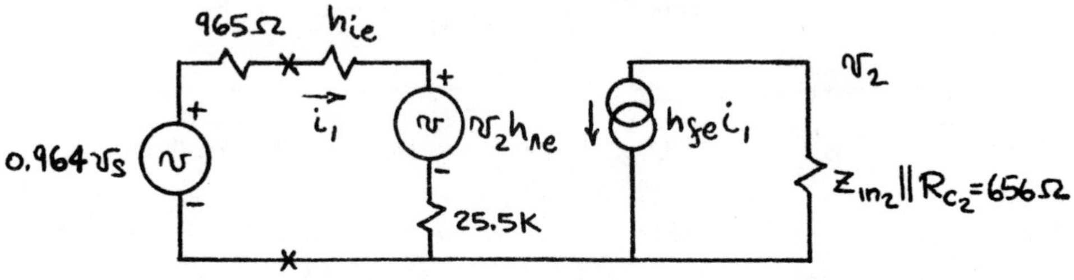

$$i_1 = \frac{0.964 v_s - h_{re} v_2}{965 + h_{ie} + 25.5K} \qquad v_2 = -h_{fe} i_1 (656)$$

$$i_1 = \frac{-v_2}{(656)(50)} = -30.5 \times 10^{-6} v_2$$

$$-30.5 \times 10^{-6} v_2 = \frac{0.964 v_s - 2\times10^{-4} v_2}{965 + 1K + 25.5K}$$

$$-838 \times 10^{-3} v_2 = 0.964 v_s - 2\times10^{-4} v_2$$

$$-v_2(838\times10^{-3}) = 0.964\, v_s \qquad \frac{v_2}{v_s} = \frac{-0.964}{838\times10^{-3}} = -1.15$$

Then the total gain $\frac{v_o}{v_s} = \frac{v_o}{v_2} \times \frac{v_2}{v_s} = (-1.15)(-116) = 134$

The gain of this amplifier could be estimated reasonably as follows:

Stage #1 has strong current-series feedback (the un-bypassed emitter resistor). Thus its gain is very nearly $-R_{L_{eff}}/R_{E_{eff}}$ where $R_{L_{eff}}$ is the 2K collector load (in parallel with h_{ie} of Q_2) and the $R_{E_{eff}}$ is the R_E plus h_{ib}. h_{ib} is approximately $h_{ie}/h_{fe} = 20\Omega$.

Thus stage #1 has an approximate gain of $\text{Gain} = \frac{-667}{500+20} = -1.28$

For stage #2 the gain is approximately $\frac{-h_{fe}R_{L_{eff}}}{h_{11}}$ where $R_{L_{eff}}$ is $5K\|5K$ or $2.5K\Omega$.

$$\text{Then Gain}\#2 \approx \frac{-(2.5K)(50)}{1K} = -125$$

Finally, the voltage divider of the input stage and the source resistance. $Z_{in}\#1 \approx h_{11} + (h_{fe}+1)R_E \approx 1K + (51)(5K) = 27K.$

The biasing network to stage #1 was found to be about 27K so that the combined parallel input impedance is about 13.5K.

Then $v_i \approx \frac{v_s(13.5K)}{13.5K+1K} = 0.93\, v_s$

Then the overall gain is estimated as follows:

$$G = \frac{v_o}{v_s} \approx 0.93(-125)(-1.28) = 149$$

This is about 11% higher than found by the more accurate procedure - well within the tolerance of the known parameters.

ELECTRONICS 3

TUNED AMPLIFIER GAIN/BANDWIDTH PROBLEM USING A JFET

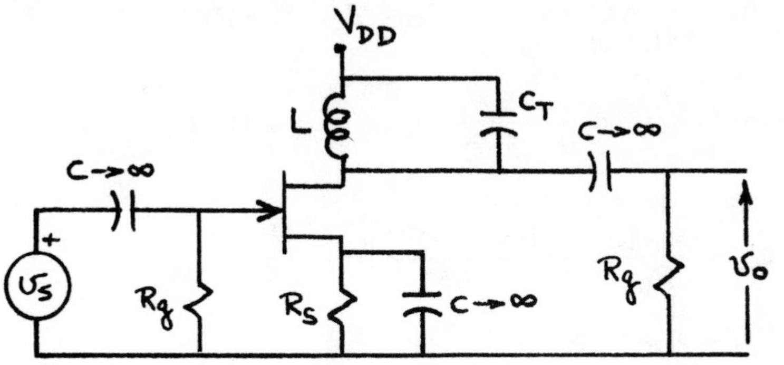

R_g = 1 megohm

L = 10μh with 1.57Ω series resistance at 5Mhz

$g_m = 2.5 \times 10^{-3}$ mhos

rd = 20KΩ (drain resistance)

Pinch off voltage is -3 volts and I_{dss} = 5 ma (I_{dss} = drain current with V_{gs} = -3 V). Select R_s for I_d = 4 ma and V_{DD} so that V_{DS} = 10 volts. Then find v_o/v_s and the 3 dB bandwidth at f_o = 5 Mhz. Also, determine the value of C_T assuming the output impedance of the JFET is real.

SOLUTION

For a JFET $$I_D \simeq I_{DS}\left(1 - \frac{|V_{GS}|}{|V_P|}\right)^2$$

Then $$I_D = 4 \cong 5\left(1 - \frac{|V_{GS}|}{3}\right)^2$$

$$\sqrt{\frac{4}{5}} = 1 - \frac{|V_{GS}|}{3}$$

$$0.89 - 1 = -\frac{|V_{GS}|}{3}, \qquad |V_{GS}| = 0.32$$

Then $R_S = \frac{0.32}{4\ ma} = 80\ \Omega$ and $V_{DD} = 10 + 0.32 = 10.32$ volts

For resonance,

$$\omega_o = \frac{1}{\sqrt{L\ C_T}}, \qquad \omega_o = 2\pi(5\times10^6)$$

$$(10\pi\times10^6)^2 = \frac{1}{10\times10^{-6}C_T} \qquad C_T = 101\ pf$$

For the coil

$$Q_o = \frac{\omega_o L}{R_S} = \frac{2\pi\times5\times10^6\times10\times10^{-6}}{1.57}$$

$$Q_o = 200 = Q_s$$

For $Q > 10$, $Q_p \simeq Q_s$ or $\frac{R_p}{\omega_o L} = \frac{\omega_o L}{R_s}$

$$R_p = Q_o\omega_o L = 200\,(2\pi\times5\times10^6\times10\times10^{-6}) = 62.8K$$

We may now draw the following equivalent circuit.

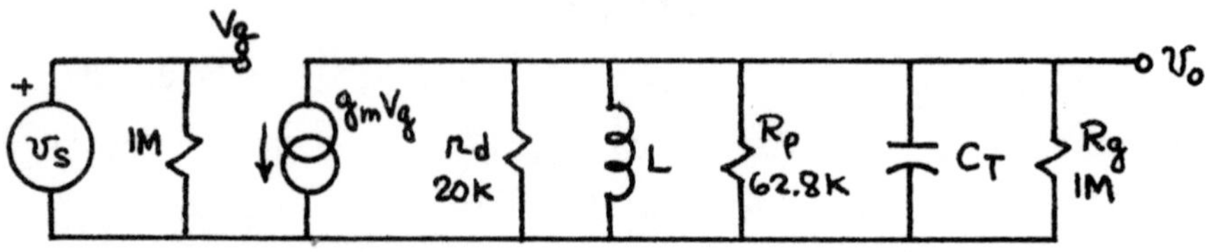

The total load seen by the drain circuit is

$$R_{L_{eff}} = r_d\,||\,R_p\,||\,R_g = \frac{1}{\frac{1}{20K}+\frac{1}{62.8K}+\frac{1}{1M}} = 14.9K\,\Omega$$

$$v_o = -g_m V_g R_{L_{eff}}$$

and since $v_g = v_s$, then $v_o/v_s = -g_m R_{L_{eff}}$

$$v_o/v_s = -2.5\times10^{-3}(14.9K) = -37.4$$

The effective Q of the drain circuit is

$$Q_{eff} = \frac{R_{L_{eff}}}{\omega_o L} = \frac{14.9K}{\omega_o L} = 47.4$$

$$3dB\ BW = \frac{\omega_o}{Q_{eff}} \Rightarrow \frac{f_o}{Q_{eff}} = \frac{5\times10^6}{47.4} = 105\ Khz$$

ELECTRONICS 4

BROADBAND AMPLIFIER USING HYBRID PI MODEL OF A BI-POLAR TRANSISTOR

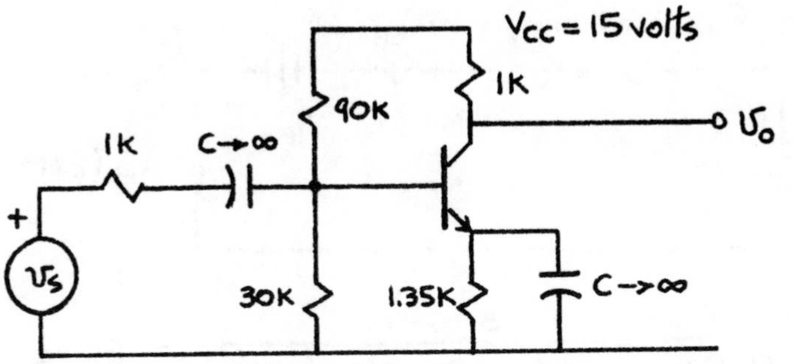

TRANSISTOR PARAMETERS:

$r_{bb'}$ (Base spreading resistance) = 200 Ω

$f(t)$ (short circuit current gain bandwidth product) = 100×10^6 hz

β (common emitter low frequency current gain ≈ h_{fe}) = 99

$C_{b'c}$ (depletion layer capacitance) = 5 p.f. at given bias point

V_{BE} (diode drop forward biased base emitter junction) = 0.6 V

Find the midband gain, v_o/v_s, the 3 dB bandwidth (assuming that the low frequency cutoff ≈ 0 hz) of this amplifier using the hybrid Pi model for the transistor.

SOLUTION

One parameter needed for the hybrid Pi model that is not given explicitly is the incremental base-emitter diode resistance rd which may be derived from the Ebers-Moll equation as

$$rd \approx \frac{0.026}{I_E}\ \Omega$$

where I_E is the dc bias value of the emitter current.

Thus we must first calculate the correct bias point for this transistor. The equivalent circuit for this calculation is shown below. We use a Thevenin equivalent for the biasing network.

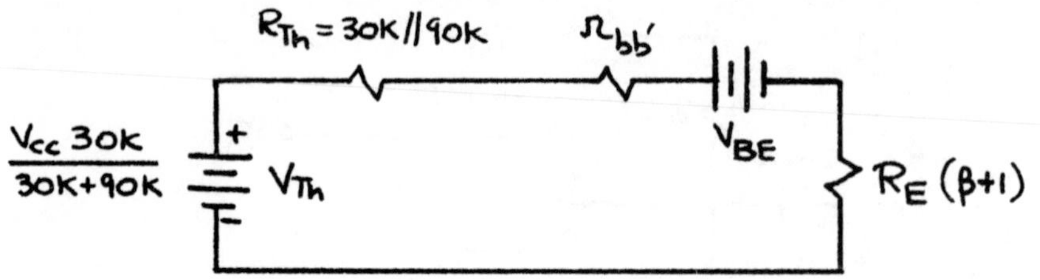

$$I_B = \frac{V_{Th} - V_{BE}}{R_{Th} + R_E(\beta+1)} = \frac{3.75 - 0.6}{22.5K + 200 + 100\,(1.35K)} = 20\,\mu a$$

Then $I_E = (\beta+1)\, I_B = 2$ ma

and $r_d = \dfrac{0.026}{2 \times 10^{-3}} = 13\,\Omega$

Now the equivalent hybrid Pi circuit:

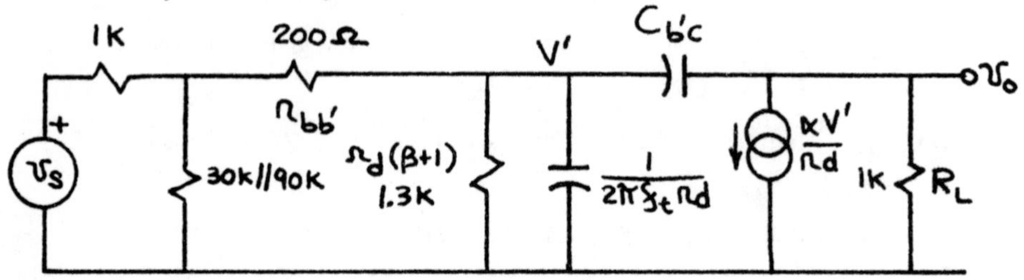

We may simplify this equivalent circuit by assuming that the bias network is large compared to Z_{in} (22.5K >> 1.5K) and assuming that R_L is small compared to the output impedance of the transistor and the impedance of $C_{b'c}$ so that all of the current from the controlled source flows through R_L. This enables us to unilaterize the circuit and replace $C_{b'c}$ with its miller value as seen by the input circuit.

$$C_{miller} \cong \left(1 + \frac{\alpha R_L}{r_d}\right) C_{b'c}$$

and if $\dfrac{\alpha R_L}{r_d} >> 1$, then $C_{miller} \cong \dfrac{\alpha R_L}{r_d} C_{b'c}$

Then the total capacitance seen at the input is

$$C_T \approx \frac{1}{2\pi f_t r_d} + \frac{\alpha R_L C_{b'c}}{r_d}$$

or $C_T \approx (1 + \alpha R_L C_{b'c}\,\omega_t)\dfrac{1}{\omega_t r_d} = 507$ p.f.

Now the equivalent circuit looks like this:

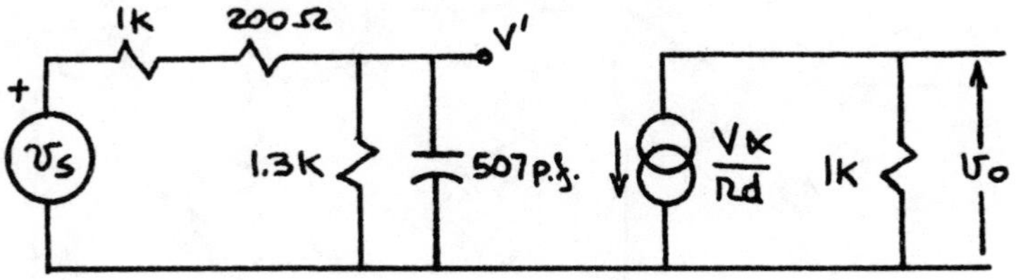

For mid band $\left|\dfrac{1}{j\omega C_T}\right| \gg r_d(\beta+1)$

$$V' = v_s \frac{(1.3K)}{1K + 200 + 1.3K} = 0.52\, v_s$$

$$v_o = \frac{-\alpha V' R_L}{r_d} \approx \frac{-0.52\, v_i(1K)}{13}$$

$$v_o/v_i = -40$$

The upper 3 dB point $\left|\dfrac{1}{\omega C_T}\right| = 1.3K \parallel 1.2K$

$$\text{or } \omega_{3dB} = \frac{1}{\tau} = \frac{1}{(507 \times 10^{-12})(1.3K \parallel 1.2K)}$$

$$\omega_{3dB} = 3.16 \times 10^6 \text{ radians/sec}$$

$$f_{3dB} = \frac{\omega_{3dB}}{2\pi} = 503 \text{ Khz}$$

ELECTRONICS 5

Calculate the power gain of the two-stage amplifier shown below. Show the gain of the two individual stages, the interstage losses and the pre-first stage losses. The ground-based "h" parameters for the transistors used in both stages are:

h_{ib} = 50 ohms; $h_{rb} = 5 \times 10^{-4}$; h_{fb} = -0.97

and $h_{ob} = 10^{-6}$ mho

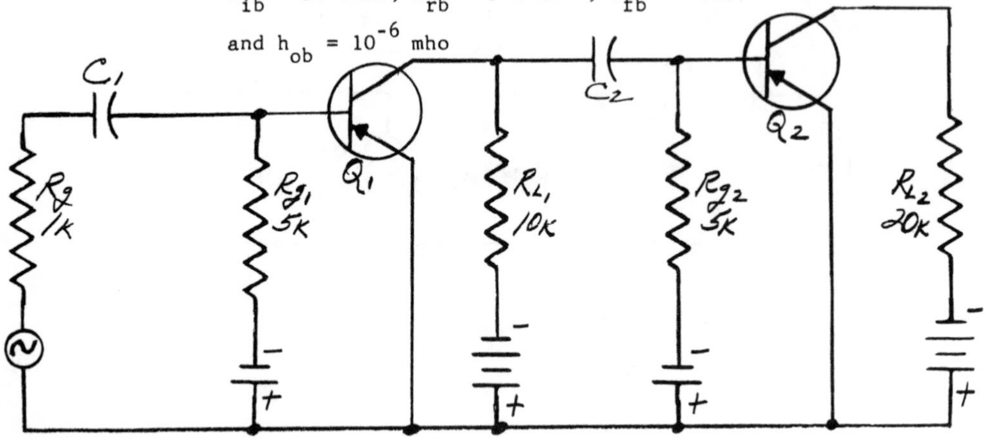

Assume reactance to be negligible.

SOLUTION

Solution based upon the following assumptions:

1. Mid-band frequency -- reactance of capacitors is negligible.
2. Power gain defined as power delivered to load divided by power delivered from source (this is power delivered to R_{L2} divided by power delivered the voltage node at the junction of C_1 and R_{g1}).
3. Power losses are defined as:
   a) Pre-first stage -- power lost in bias resistor R_{g1}.
   b) Inter-stage -- power lost in R_{L1} and R_{g2}.
   c) Power losses -- at signal frequency only, d.c. bias losses not considered.

Parameters are given in common base configuration; since transistors are operated in common emitter orientation, the parameters must be converted to common emitter form. Also, gain and impedance equations must either be derived or found in a transistor handbook.

Handbook conversion tables (from Fig. 4.11, G.E. Transistor Manual, 6th ed.) give the following relationships:

$$h_{ie} = \frac{h_{ib}}{1+h_{fb}} = 1670\ \Omega \qquad h_{re} = \frac{h_{ib}h_{ob}}{1+h_{fd}} = 11.7 \times 10^{-4}$$

$$h_{fe} = \frac{-h_{fd}}{1+h_{fb}} = 32 \qquad h_{oe} = \frac{h_{ob}}{1+h_{fd}} = 3.3 \times 10^{-5}\ \mho$$

And (from Fig. 4.14):

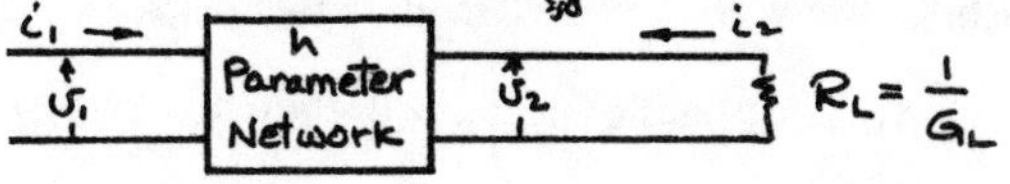

$$\text{I.} \quad Z_{in} \triangleq \frac{v_1}{i_1} = \frac{h_{ie}h_{oe} - h_{fe}h_{re} + h_{ie}G_L}{h_{oe} + G_L} = h_{ie} - \frac{h_{re}h_{fe}}{h_{oe}+G_L}$$

$$\text{II.} \quad A_v \triangleq \frac{v_2}{v_1} = \frac{-h_{fe}R_L}{h_{ie} + R_L(h_{ie}h_{oe} - h_{fe}h_{re})} = \frac{-h_{fe}}{h_{ie}G_L + (h_{ie}h_{oe} - h_{re}h_{fe})}$$

The gain of each stage may be calculated by breaking the circuit as follows:

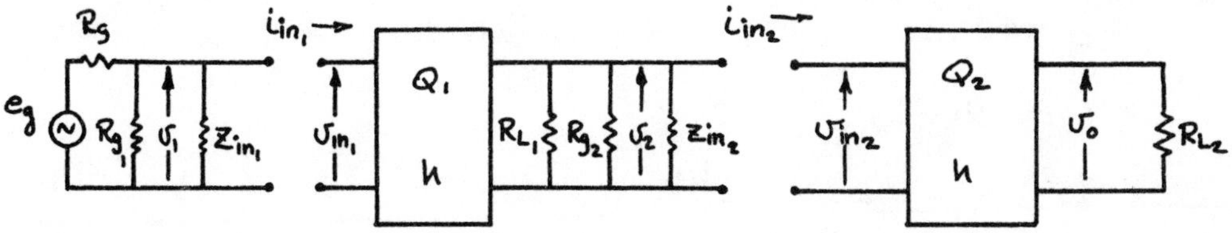

Definitions:

v_1 = Voltage at base of Q_1

$v_1 = v_{in_1}$

v_2 = Voltage at collector of Q_1

$v_2 = v_{in_2}$

v_o = Voltage across load R_{L_2}

$$v_1 = e_g \frac{(R_{g_1} \text{ in parallel with } Z_{in_1})}{R_{g_1} + (R_{g_1} \text{ in parallel with } Z_{in_1})}$$

Calculation of stage Q_2 (Eqn II):

$$\frac{v_o}{v_{in_2}} = A_v = \frac{-h_{fe}}{h_{ie} G_L + (h_{ie} h_{oe} - h_{oe} h_{fe})}$$

$$= \frac{-32}{(1670)(0.5 \times 10^{-4}) + (1670)(3.3 \times 10^{-5}) - (11.7 \times 10^{-4})(32)} = -320$$

To calculate gain of first stage, the impedance Z_{in_2} must first be obtained; this appears in parallel with R_{L_1} and R_{g_2}, all of which appear as a parallel load for stage Q_1. Then Z_{in_2} (from Eqn I) is given as:

$$Z_{in_2} = h_{ie} - \frac{h_{re} h_{fe}}{h_{oe} + G_L} = 1670 - 450 = 1220\ \Omega$$

Then the effective load for Q_1 is:

$$G'_{L_1} = \frac{1}{R'_{L_1}} = \frac{1}{R_{g_2}} + \frac{1}{R_{L_1}} + \frac{1}{Z_{in_2}} = \frac{1}{5K} + \frac{1}{10K} + \frac{1}{1.22K} = \frac{1}{895} = 1.12 \times 10^{-3}\ \mho$$

$$R'_{L_1} = 895\ \Omega$$

Then (from Eqn II):

$$\frac{V_2}{V_{in_1}} = A_v = \frac{-h_{fe}}{h_{ie} G'_L + (h_{ie} h_{oe} - h_{oe} h_{fe})}$$

$$= \frac{-32}{(1670)(1.12 \times 10^{-3}) + (1670)(3.3 \times 10^{-5}) - (11.7 \times 10^{-4})(32)} = -16.9$$

The total voltage gain from v_1 to v_o is:

$$\frac{v_o}{v_1} = \left(\frac{v_{in_2}}{v_1}\right)\left(\frac{v_o}{v_{in_2}}\right) = (-320)(-16.9) = 5400$$

To find the power delivered by the source, Z_{in_1} must be calculated (Eqn I):

$$Z_{in_1} = \frac{V_{in}}{i_{in_1}} = h_{ie} - \frac{h_{re} h_{fe}}{h_{oe} + G'_{L_1}}$$

$$= 1670 - \frac{(11.7 \times 10^{-4})(32)}{(3.3 \times 10^{-5}) + (1.12 \times 10^{-3})} = 1670 - 30 = 1640\ \Omega$$

Then the power delivered by the generator is:

$$P_{in} = \frac{V_1^2}{\text{Parallel combination of } R_{g_1} \text{ and } Z_{in_1}} = \frac{v_1^2}{\frac{(5\times10^3)(1640)}{5\times10^3 + 1640}} = \frac{v_1^2}{1.25\times10^3}$$

and the power delivered to the load is:

$$P_o = \frac{v_o^2}{R_{L_2}} = \frac{v_o^2}{20\times10^3}$$

Then the power gain is:

$$G = \frac{P_o}{P_{in}} = \frac{v_o^2/20\times10^3}{v_1^2/1.25\times10^3} = 0.0625\left(\frac{v_o}{v_1}\right)^2 = 0.0625(5400)^2$$

$$= 1.82\times10^6$$

$$G_{db} = 10 \log G = 62.6 \text{ db}$$

Now consider the inter-stage power loss (lost in parallel combination of R_{L_1} and R_{g_2}):

$$\frac{(R_{L_1})(R_{g_2})}{R_{L_1} + R_{g_2}} = \frac{(10\times10^3)(5\times10^3)}{10\times10^3 + 5\times10^3} = 3.33\times10^3\,\Omega$$

$$P_{I.L.} = \frac{v_2^2}{3.33\times10^3} = \frac{(16.9\,v_1)^2}{3.33\times10^3}$$

The total loss is then:

$$P_{Losses} = \frac{v_1^2}{5\times10^3} + \frac{(16.9\,v_1)^2}{3.33\times10^3}$$

But $v_1 = 0.556\, e_g$

$$P_{Losses} = \frac{(0.556 e_g)^2}{5\times10^3} + \frac{[(0.556)(16.9) e_g]^2}{3.33\times10^3}$$

$$= 27.7\times10^{-3} e_g^2$$

ELECTRONICS 6

Subject: Electronics: Transistor circuit

Consider the following schematic diagram:

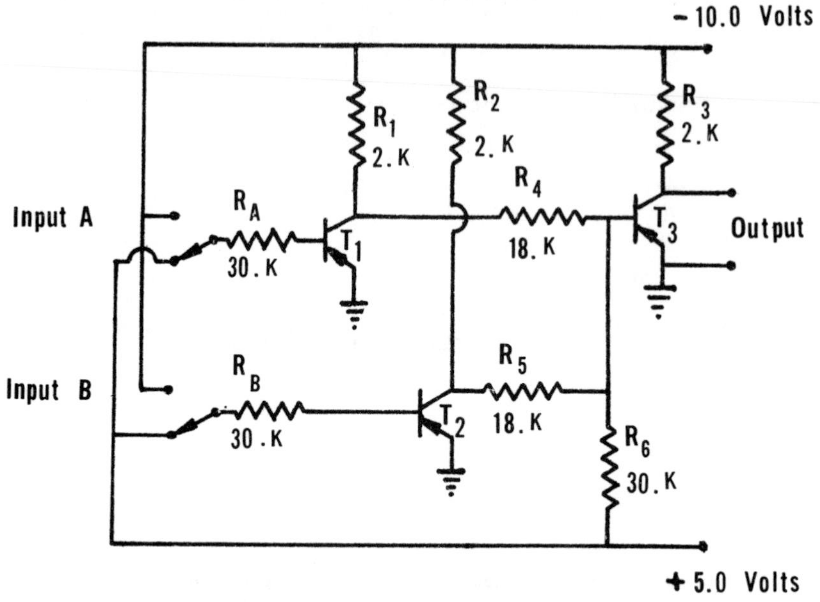

For the schematic diagram shown assume the following:

β (min) for transistors T_1 and T_2	20.
I_{co} (max) for any transistor	100. μa
V_{ce} (saturated) for any transistor (magnitude)	0.1 volt
V_{be} (max) for any "ON" transistor (magnitude)	0.35 volt
Resistor tolerance (max) from nominal values shown	± 10.%
Power Supplies	Constant Voltage

In the above schematic each input (A or B) may be switched to either the +5.0 volt or -10.0 volt supply.

For the WORST CASE conditions (all values at tolerance limit) determine the following two values:

(1) Wt. 7 What minimum β must transistor T_3 have for the output to be -0.1 volt when the inputs A and B are not alike?

(2) Wt. 3 What will be the minimum bias voltage on transistor T_3 when it is cut off?

SOLUTION

(1) Assume switch "A" up and "B" down, then T_1 may be saturated and T_3 will be saturated; then replace all resistor values (by either plus or minus 10% to give the worst case) and T_2 by an equivalent current source equal to the cutoff leakage current of I_{co}:

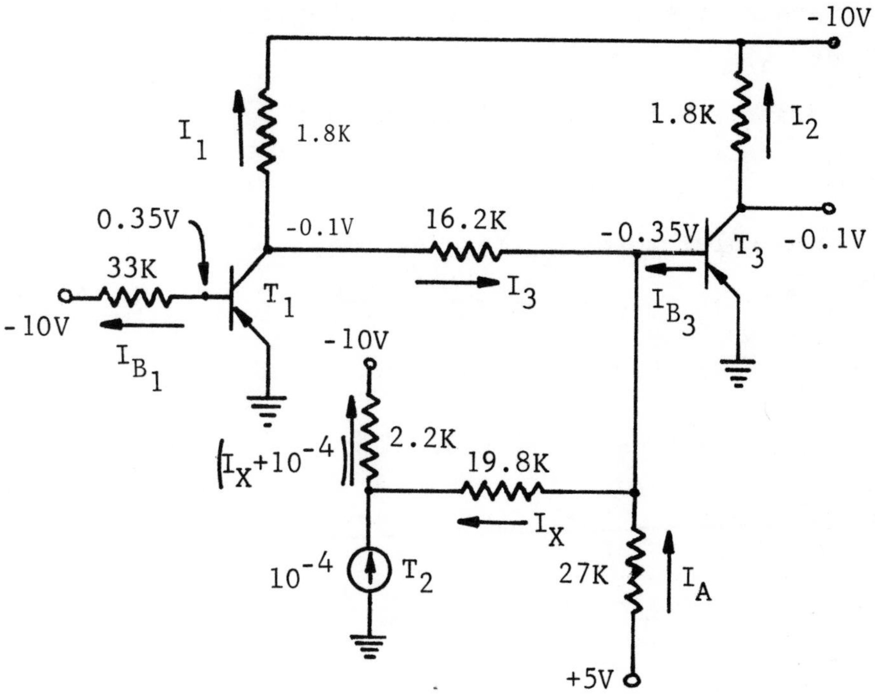

Then:

$$I_1 = \frac{10 - 0.1}{1.8 \times 10^3} = 5.5 \times 10^{-3} = I_2 \text{ (if saturated)}$$

$$I_{B_1} = \frac{10 - 0.35}{33 \times 10^3} = 0.292 \times 10^{-3}$$

β_1(minimum) = 20, so $I_{B_1}\beta_1 = (20)(0.29 \times 10^{-3})$
$= 5.84 \times 10^{-3}$

$\therefore I_{B_1} \beta_{1_{min}} > 5.5$, thus T_1 is indeed saturated.

$$I_A = \frac{5.35}{27 \times 10^3} = 0.198 \times 10^{-3}$$

Then I_X may be found from Kirchhoff's law, (around the voltage loop starting at the base of T_3):

$$+0.35 + 19.8 \times 10^3 I_X + 2.2 \times 10^3(10^{-4} + I_X) - 10 = 0$$

$$I_X = 0.429 \times 10^{-3}$$

Then solving for T_3 base current:

$$I_{B_3} = -I_3 + I_X - I_A = -\frac{0.35 - 0.1}{16.2 \times 10^3} + 0.429 \times 10^{-3} - 0.198 \times 10^{-3}$$

$$= -\frac{(0.35 - 0.1)}{16.2 \times 10^3} + 0.429 \times 10^{-3} - 0.198 \times 10^{-3}$$

$$= 0.215 \times 10^{-3}$$

$$\therefore \beta_{3_{min}} = \frac{5.5 \times 10^{-3}}{0.215 \times 10^{-3}} = 25.6$$

(2) Consider both switches to be "up", then:

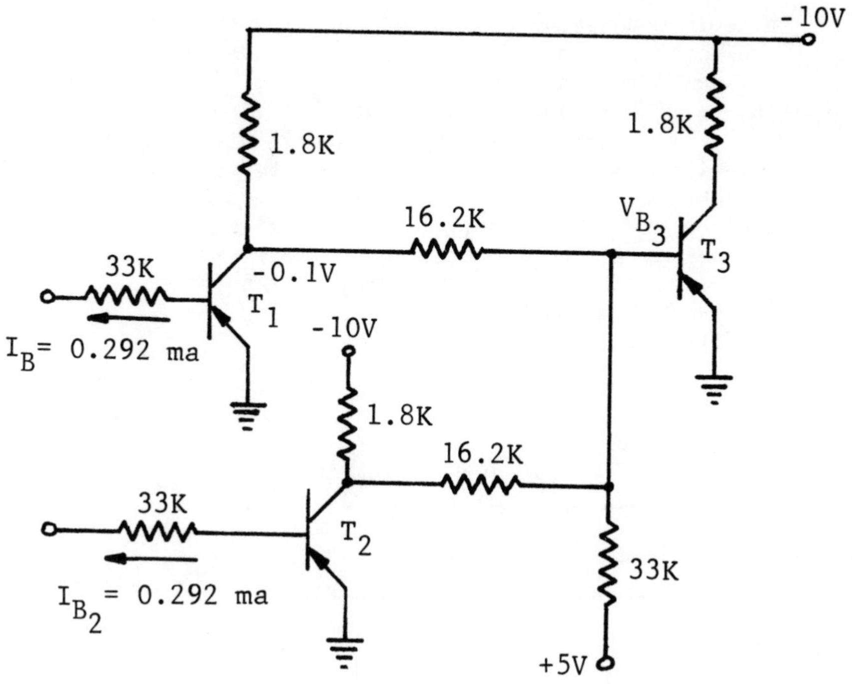

Then for V_{B_3}:

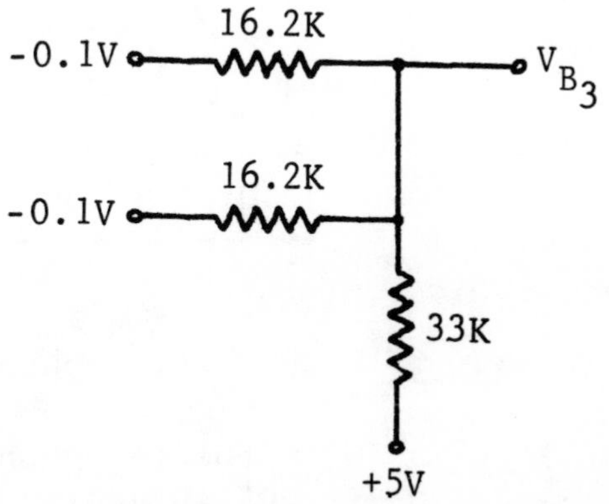

$$V_{B_3} = 5 - \frac{5.1 \text{ x } 33}{8.1 + 33} = +0.9 \text{ volts (min. bias)}$$

ELECTRONICS 7

COMMUNICATIONS: TELEPHONE CIRCUIT

In the figure below, the simple low pass filter network consisting of the balanced inductance (L) and the capacitor (C) is used to provide isolation between a telephone set bridged on the wire line and the carrier frequency equipment operating on the same wire line. The designed cut-off frequency of this low pass filter is 3.0 kcs. Determine and sketch the impedance as seen by the wire line and looking from the wire line into the filter network as a function of frequency over the voice range for a telephone set having an impedance of 600 ohms when connected or in use under the conditions:

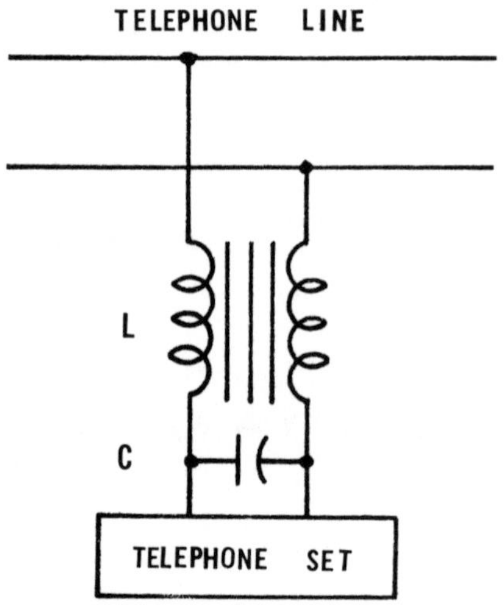

(1) Wt. 5 With the telephone set connected or in use.

(2) Wt. 5 With the telephone set disconnected or not in use.

SOLUTION

The question arises, are the two inductances to have coupling between them or not? Since the statement is made, a "simple low pass filter ..", any mutual inductance between the L's is not simple.

Proceeding on the basis of M = 0

Balanced circuit:

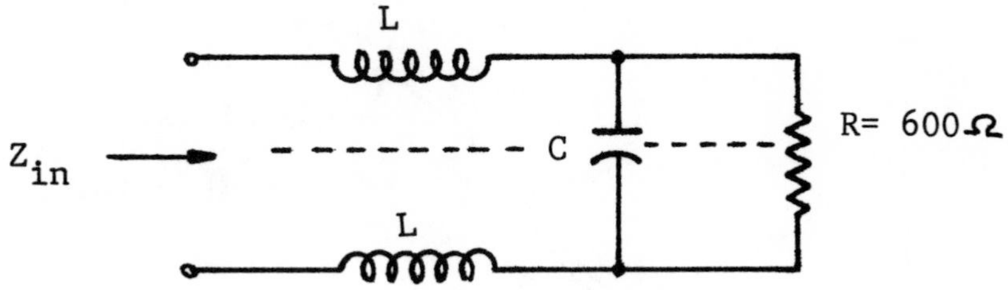

Unbalanced equivalent:

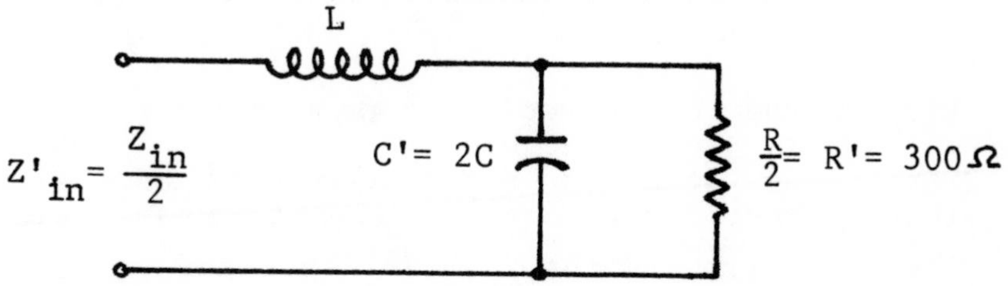

By Laplace Transformation:

$$Z'_{in} = sL + \frac{R'}{1 + sR'C'}$$

$$= \frac{(s + a)(s + b)}{(s + c)}$$

where $a = \frac{1}{2R'C'} - \frac{1}{4R'^2C'^2} - \frac{1}{LC'}$

$b = \frac{1}{2R'C'} + \frac{1}{4R'^2C'^2} - \frac{1}{LC'}$

$c = \frac{1}{R'C}$ note $c > b > a$

we assume a and b are real and $R^2 < \frac{L}{4C'}$

Using the method of asymptotes, Z'_{in} may be plotted for $s(j\omega)$ very, very low

$Z'_{in} = L\frac{as}{c} = R'$ $\omega = 2\pi f$ $j = \sqrt{-1}$

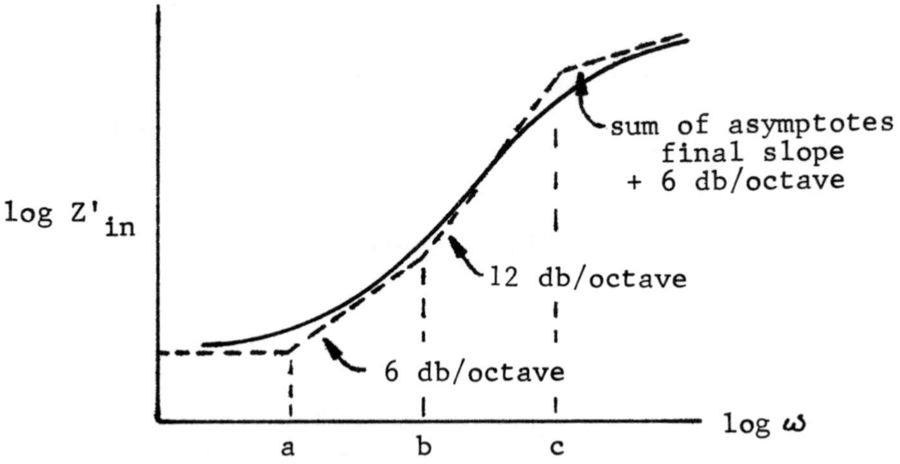

First corner occurs at s = a Asymptote has + slope 6 db/octave.
Next corner occurs at s = b Asymptote has + slope
Third corner occurs at s = c Asymptote has - slope

The solid line is then log Z'_{in} plotted vs. log ω

The departure of Z'_{in} at "corners" is 3 db.

Since cut-off frequency is stated as 3 KC,

"a" has a value of $2\pi \times 3 \times 10^3 = 6\pi \times 10^3$ rad/sec.

Now $Z'_{in} = 1/2\ Z_{in}$; also by proper choice of L and C the corners a and b may be moved closer together.

Suppose $R^2 = \frac{L}{4C}$

then $a = b = \frac{c}{2}$ and $Z'_{in} \equiv \frac{(s + a)^2}{s + c}$

In this case the first "corner" occurs at $a = \omega$ and the asymptotic slope between a and c is 12 db/octave.

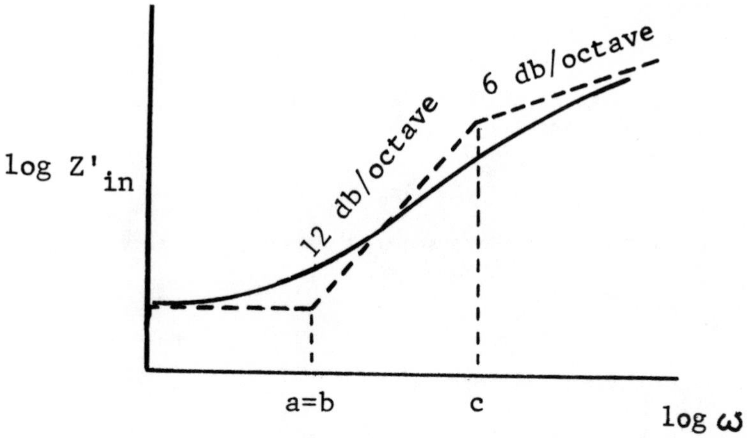

(2) If the telephone set is disconnected (purely reactive network):

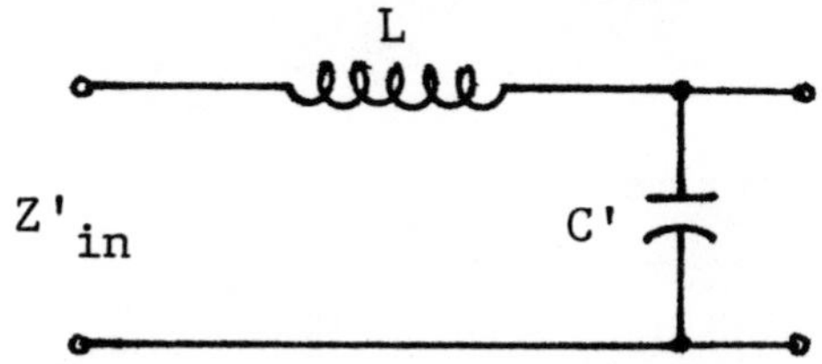

$$Z'_{in} = j\omega L + \frac{1}{j\omega C'}$$

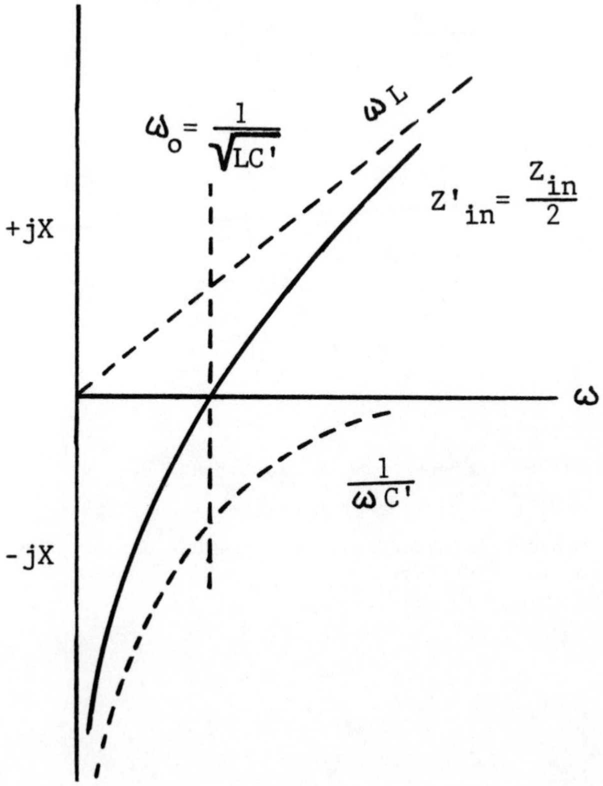

ELECTRONICS 8

In the modulating circuit sketched below, the modulating signal (see spectrum sketch) is limited to angular frequencies ω where

$$\omega_{m_1} < \omega < \omega_{m_2} \lll \omega_c.$$

Where ω_m = modulating frequency; and ω_c = carrier frequency.

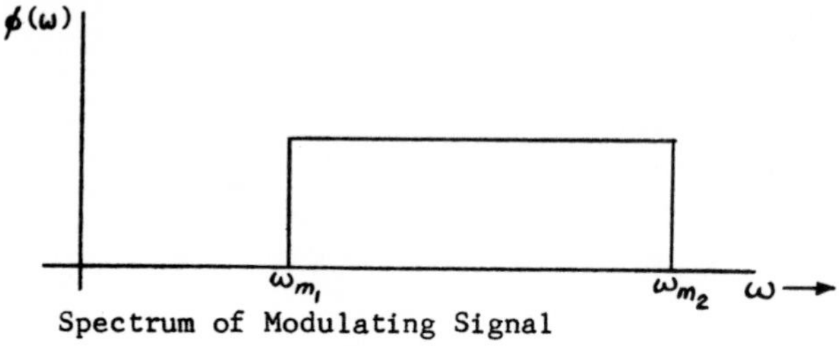

Spectrum of Modulating Signal

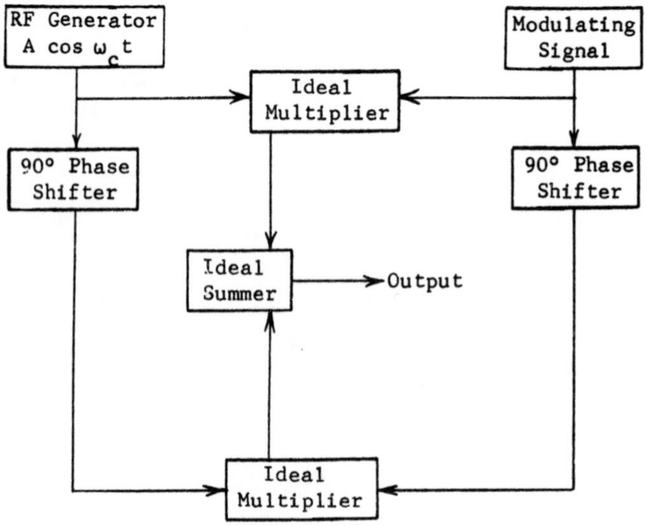

REQUIRED:

Wt.

5 (a) Sketch the spectrum of the output.

2 (b) What is this type of modulated signal called?

3 (c) Sketch another circuit which would produce the same output spectrum.

SOLUTION

(a) The in-phase input is as follows:

$A \cos \omega_c t \cdot \phi \cos \omega_m t =$

$\frac{A\phi}{2}\left[\cos(\omega_c - \omega_m)t - \cos(\omega_c + \omega_m)t\right]$

The 90° phase shift is as follows:

$A \sin \omega_c t \cdot \phi \sin \omega_m t =$

$\frac{A\phi}{2}\left[\cos(\omega_c - \omega_m)t + \cos(\omega_c + \omega_m)t\right]$

The sum of above two expressions is:

$A\phi \cos(\omega_c - \omega_m)t$

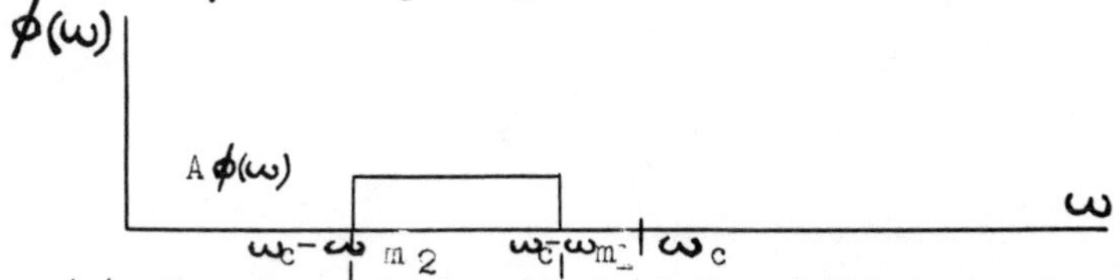

(b) This is a single side band AM modulated signal.

(c) A balanced modulator and sideband filter would produce the same output spectrum:

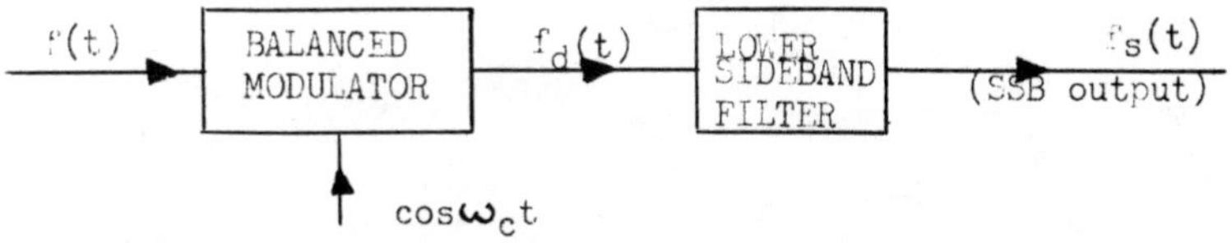

Reference: Information Transmission, Modulation and Noise by Schwartz, McGraw-Hill p. 106, 107.

ELECTRONICS 9

A certain transmitter has an effective radiated power of 9 KW with the carrier unmodulated and 10.125 KW when the carrier is modulated by a sinusoidal signal.

REQUIRED:

Wt.

5 (a) Determine the percent modulation at 10.125 KW output.

5 (b) Determine the total effective radiated power if in addition to the sinusoidal signal, the carrier is simultaneously modulated 40% by an audio wave.

SOLUTION

(a) Assume transmission is through a 1 ohm resistance. For the unmodulated carrier with carrier frequency ω_o:

$$v_o(t) = A \cos\omega_o t$$

$$P = \int v_o(t) = \frac{A^2}{2} = 9 \text{ KW, since:}$$

$$P = \frac{1}{2\pi}\int_0^{2} \frac{v^2(t)}{R}\,dt = \frac{A^2}{2\pi}\int_0^{2\pi} \cos^2\omega_o t\,dt$$

$$= \frac{A^2}{2\pi}\int_0^{2} \frac{\cos 2\omega_o t + 1}{2}\,dt = \frac{A^2}{2\pi}\cdot\frac{2\pi}{2} = \frac{A^2}{2}$$

Then, for the modulated carrier with modulating frequency ω_1, and modulating coefficient "m", we obtain

$$v_o(t) = A\,(1 + m \cos\omega_1 t) \cos\omega_o t$$

$$= A \cos\omega_o t + \frac{Am}{2}\cos(\omega_o + \omega_1)t + \frac{Am}{2}\cos(\omega_o - \omega_1)t$$

since a basic trigonometric relation states:

$$\cos\alpha \cos\beta = \frac{1}{2}\cos(\alpha + \beta) + \frac{1}{2}\cos(\alpha - \beta)$$

Therefore: $P = \dfrac{A^2 + \left(\frac{Am}{2}\right)^2 + \left(\frac{Am}{2}\right)^2}{2} = 9\left[(1 + 2\,\frac{m}{2})^2\right] = 10.125$

$\frac{m}{2} = \sqrt{\frac{1.125}{18}} = 0.25$ and m = 0.5, or 50% ANS.

(b) Denoting the audio wave frequency with ω_2, we obtain:

$$v_o(t) = A\,(1 + 0.5 \cos\omega_1 t + 0.4 \cos\omega_2 t) \cos\omega_o t$$

$$= A\cos\omega_o t + \frac{0.5A}{2}\cos(\omega_o + \omega_1)t + \frac{0.5A}{2}\cos(\omega_o - \omega_1)t$$

$$+ \frac{0.4A}{2}\cos(\omega_o + \omega_2)t + \frac{0.4A}{2}\cos(\omega_o - \omega_2)t$$

Then the effective radiated power is:

$$P = \frac{A^2 + 2\left(\frac{0.5A}{2}\right)^2 + \left(\frac{0.4A}{2}\right)^2}{2} = 10.125 + 9\left[2\left(\frac{0.4}{2}\right)^2\right]$$

$$= 10.125 + 0.72 = 10.845 \text{ KW}$$ ANS.

The frequency spectrum is as follows:

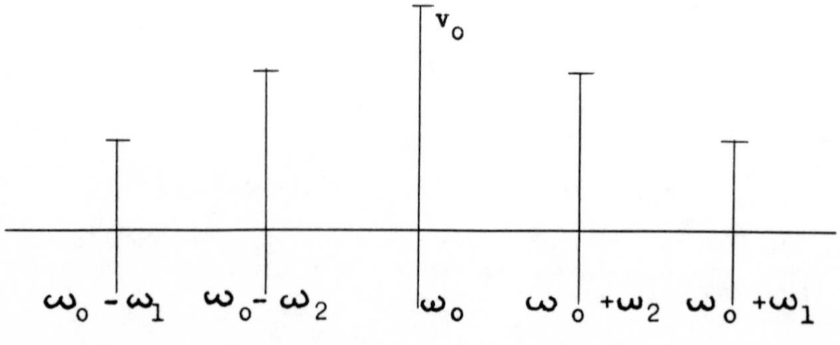

ELECTRONICS 10

INFORMATION THEORY

In order to specify the communications link for a closed circuit television system, the bit rate must be known.

The monochrome television picture signal of this system requires 10 distinct levels of brightness for good resolution. This television system also includes the following parameters:

(1) Frame rate, 15 frames per second
(2) Lines per frame, 1200
(3) Discrete picture elements, 100 per line

REQUIRED:

Determine the channel capacity in bits per second required to transmit the above video signal with all levels equally probable and with all elements assumed to vary independently.

List any assumptions that you make.

SOLUTION

The number of different possible pictures is

$$P = 10 \times 10 \ldots \times 10 = 10^{1200 \times 100} = 10^{1.2 \times 10^5}$$

Probability of each element or picture is

$$= \frac{1}{10^{1.2 \times 10^5}}$$

The Channel Capacity is defined as

$$C = \lim_{T \to \infty} \frac{1}{T} \log_2 M(T)$$ where M(T) is the total number of messages in T seconds.

$$= S \log_2 P \quad \text{bit/sec}$$ where S is the signalling speed.

$S = 15$

$$C = 15 \log_2 10^{1.2 \times 10^5}$$

$$= 1.8 \times 10^6 \log_2 10$$

$$= 6.0 \times 10^6 \text{ bit/second}$$

Assumption: It is assumed that the signal to noise ratio is large or the error probability is small so the channel is deemed noiseless.

ELECTRONICS 11

The circuit shown below is an N-channel Junction Field-Effect Transistor with self-bias and a pinch-off voltage of -3 volts. At that value of pinch-off voltage, the current is 6 mA. The breakdown voltage for this transistor is 30 volts.

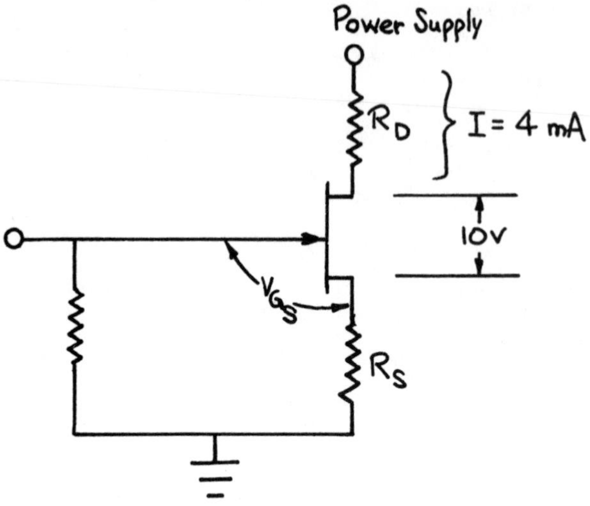

REQUIRED: Design the above circuit so that the device will be biased at approximately 10 V drain-to-source and have a channel current of approximately 4 mA.

SOLUTION

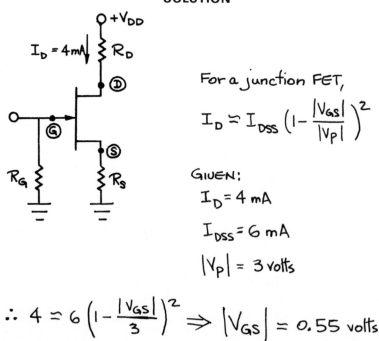

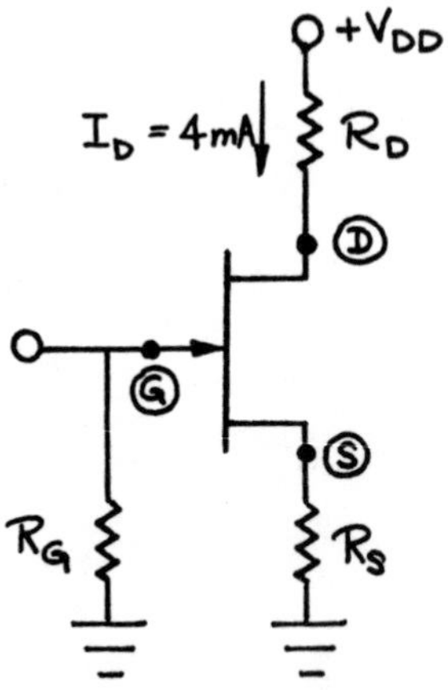

For a junction FET,

$$I_D \simeq I_{DSS}\left(1 - \frac{|V_{GS}|}{|V_P|}\right)^2$$

GIVEN:

$I_D = 4$ mA

$I_{DSS} = 6$ mA

$|V_P| = 3$ volts

$$\therefore\ 4 \simeq 6\left(1 - \frac{|V_{GS}|}{3}\right)^2 \Rightarrow |V_{GS}| = 0.55 \text{ volts}$$

Gate leakage current I_{GSS} is typically of the order of nanoamperes. Choose $R_G = 1\,M\Omega$, so that the gate remains within millivolts of ground potential. Choose R_S to obtain the proper V_{GS}.

$$|V_{GS}| = 0.55 \text{ volts},\ I_D = 4\,mA \Rightarrow R_S = \frac{0.55}{4\times10^{-3}} \simeq 140\,\Omega$$

Since the breakdown voltage is 30 volts, choose $V_{DD} = 24$ volts.

$$V_{DS} = V_{DD} - I_D(R_D + R_S)$$

$$10 = 24 - 4\times10^{-3}(R_D + R_S)$$

$$R_D + R_S = \frac{14}{4\times10^{-3}} = 3500\,\Omega$$

$$\therefore R_D = 3500 - 140 = 3360\,\Omega$$

If 10% resistors are used, the available values are:

$R_S = 150\,\Omega$ and $R_D = 3.3\,K\Omega$

ELECTRONICS 12

A series of remote stations is being planned to feed data to a large computer. The data are to be sent by the remote stations and recorded on a tape recording unit and then fed to the computer as necessary.

The data are expected to arrive from the remote stations Poisson distributed at an average rate of 10 transmissions from remote stations per hour. The recording time of the data varies exponentially, with a mean time of four minutes.

REQUIRED:

(a) What is the average waiting time for a remote station before the data will begin to record?

(b) A second tape unit including automatic switching equipment is available at a cost of \$2.50 per hour. The telephone lines cost 4¢ a minute per line when used. Is the second unit economically warranted? Show sufficient calculations to justify your answer.

SOLUTION

(a) This is the single stage, single server queueing model with Poisson arrivals and exponential service. Any basic operations research text gives the desired queue equations (e.g. Sasieni, Yaspan, and Friedman: Operations Research - Methods and Problems, John Wiley, p126-138).

mean arrival rate λ = 10 transmissions/hour

mean service time = $1/\mu$ = 4 minutes = 1/15 hour

mean service rate μ = 15 recordings/hour

Average waiting time of an arrival

$$E(w) = \frac{\lambda}{\mu(\mu - \lambda)} = \frac{10}{15(15 - 10)} = \frac{10}{75} \text{ hour} = 8 \text{ minutes}$$

(b) This is the single stage, two server queueing situation. The various expectations for this model may also be obtained from a basic operations research (or queueing) text.

Before proceeding, a quick check can be made. We know a second recorder will substantially reduce (but not eliminate) waiting time. If the elimination of waiting time would not justify the second recorder then we need not bother to make the exact computation. Instead, we could simply conclude the second recorder is not economically warranted.

hourly saving (assuming elimination of waiting time)
= mean arrival rate x mean waiting time reduction x line charge

$$= \lambda[E(w_1) - E(w_2)] \times 0.04 = (10)(8 - 0)(0.04) = \$3.20$$

hourly cost = \$2.50

Thus we have been unable to show that the second recorder is uneconomical at zero waiting time. We must, therefore, proceed to compute the expectation of average waiting time.

$$P_o = \frac{1}{\left[\sum_{n=0}^{k-1} \frac{1}{n!}\left(\frac{\lambda}{\mu}\right)^n\right] + \frac{1}{k!}\left(\frac{\lambda}{\mu}\right)^k \frac{k\mu}{k\mu - \lambda}}$$

For two service facilities:

$$k = 2 \qquad \lambda = 10 \qquad \mu = 15$$

$$P_o = \frac{1}{\frac{\lambda}{\mu} + \frac{1}{2}\left(\frac{\lambda}{\mu}\right)^2 \frac{2\mu}{2\mu-\lambda}} = \frac{1}{\frac{10}{15} + \frac{100}{450}\cdot\frac{30}{20}} = \frac{1}{1} = 1$$

Average waiting time of an arrival

$$E(w) = \frac{\mu\left(\frac{\lambda}{\mu}\right)^k}{(k-1)!(k\mu - \lambda)^2} P_o = \frac{15\cdot\left(\frac{10}{15}\right)^2}{(30 - 10)^2}(1) = \frac{1}{60} \text{ hour}$$

$$= 1 \text{ minute}$$

hourly saving of second recorder $= \lambda[E(w_1) - E(w_2)] \times 0.04$

$$= (10)(8.0 - 1.0)(0.04) = \$2.80$$

Thus the second recorder is economically warranted, for the hourly saving exceeds the hourly cost.

7

Control Systems

In the past, most control system problems were classic in nature (i.e., method of solution by root locus, Bode and Nyquist plots, etc.); future problems may include both state variable conversions and discrete systems involving the z-transform. Although, in this author's opinion, the bulk of most problems will probably continue to be in the classical format, a very short review of state variable conversion will be presented along with an introductory review of discrete (sampled data) systems.

Control is concerned with regulation or control of the output (i.e., output shaft position, velocity, temperature, etc.) in comparison to the input reference command. The analysis usually involves three areas:

PROBLEM FORMULATION. This involves making a mathematical model of the various parameters of the system. Usually first finding the differential equation of the open loop system, then converting to Laplace transforms and making the block diagram representation (one can usually assume all initial conditions to be zero).

DETERMINING SYSTEM STABILITY. Closed loop stability for linear systems (and fortunately to date almost all of the control systems problems have been of the linear kind) and usually can be found by the Routh-Hurwitz technique of analysis.

SYSTEM PERFORMANCE. One aspect of system performance normally considered is system error (after a long period of time) for various inputs (i.e., unit step, ramp, etc.). Also involved in performance analysis is percent overshoot, time-to-the-first-peak, etc., as the response to a step input (usually this is compared to second order system response) as some parameter, such as gain, is changed. Root locus is usually the preferred method for this type of response. On the other hand, if system performance is concerned with frequency response (i.e., driven resonant frequency, bandwidth, phase and/or gain margin) then Bode plots or Nichol's charts are usually considered.

Consider the following electro-mechanical system.

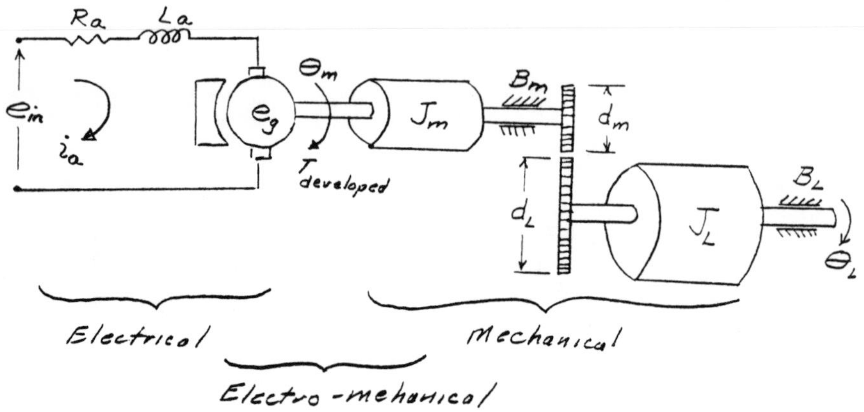

For the electrical portion one would simply use Kirchhoff's Law as $\Sigma v's = 0$:

$$e_{in} = v_R + v_L + e_g$$

where e_g = generated voltage (i.e., induced voltage) proportional to strength of the magnetic field (here, considered to be a permanent magnetic, or constant field) and armature velocity, therefore

$$e_g = K_g \omega = K_g \frac{d\Theta}{dt}$$

thus:

$$e_{in} = i_a R_a + L_a \frac{di_a}{dt} + K_g \frac{d\Theta_m}{dt} \qquad (1)$$

Also consider the mechanical portion of the circuit. Any torque developed by the armature is "used" to drive the load; here one simply sums torques $\Sigma T's = 0$:

$$T_{\substack{developed \\ by\ armature}} = J_m \frac{d^2\Theta_m}{dt^2} + B_m \frac{d\Theta_m}{dt} + T_{load}$$

But the load torque as "seen" by the motor shaft, Θ_m is reflected through the gear ratio as:

$$T_L = \eta^2 J_L \frac{d^2\Theta_m}{dt^2} + \eta^2 B_L \frac{d\Theta_m}{dt}, \qquad \eta = \frac{d_m}{d_L}$$

$$\therefore T_{developed} = \underbrace{(J_m + \eta^2 J_L)}_{J_{effective}} \frac{d^2\Theta_m}{dt^2} + \underbrace{(B_m + \eta^2 B_L)}_{B_{effective}} \frac{d\Theta_m}{dt} \tag{2}$$

Now the electromechanical relationship is such that the developed torque is proportional to magnetic flux (recall that the field is constant) and to the armature current, therefore

$$T_{developed} = K_m i_a \tag{3}$$

The three equations of interest are again repeated below except that at this point, the manipulations become much easier if we convert to Laplace Transforms (and making the simplifying assumption that all initial conditions are zero, then

$$\frac{d\Theta}{dt} \Rightarrow S\Theta, \quad \frac{d^2\Theta}{dt^2} \Rightarrow S^2\Theta, \text{ etc.})$$

(1) $E_{in} = I_a R_a + L_a S I_a + K_g S \Theta_m$

(2) $T_{developed} = J_e S^2 \Theta_m + B_e S \Theta_m$

(3) $T_{developed} = K_m I_a$

Combining equations (2) and (3) and solving for I_a, then substituting into equation (1), we obtain:

(1) $E_{in} = (R_a + SL_a) I_a + K_g S \Theta_m$

(2) & (3) $I_a = \frac{1}{K_m}(J_e S^2 + B_e S)\Theta_m$

then

$$E_{in} = (R_a + SL_a)\frac{1}{K_m}(J_e S^2 + B_e S)\Theta_m + K_g S \Theta_m$$

The inductive term is usually small (i.e., the electric time constant is usually **much** smaller than the mechanical time constant), so that if L_a is neglected, then:

$$E_{in} = \frac{R_a}{K_m}(J_e S^2 + B_e S)\Theta_m + K_g S \Theta_m$$

$$= \left(\frac{R_a J_e}{K_m} S^2 + \underbrace{\frac{R_a B_e}{K_m} S + K_g S}_{\text{combining coefficients}}\right)\Theta_m$$

$$K_B = \frac{R_a B_e}{K_m} + K_g$$

$$E_{in} = \left(\frac{R_a J_e}{K_m} S^2 + K_B S\right)\Theta_m$$

$$= K_B\left[\left(\frac{R_a J_e}{K_m K_B}\right)S + 1\right]S\Theta$$

And since the units of S are rad/sec, then $\frac{R_a J_e}{K_m K_B} = \tau$ (seconds), an electromechanical time constant.

$$E_{in} = K_B(S\tau + 1)S\Theta_m$$

or the transfer function model then becomes:

$$\frac{\Theta_m}{E_{IN}} = \frac{1/K_B}{S(S\tau + 1)} = \frac{K}{S(S\tau + 1)}$$

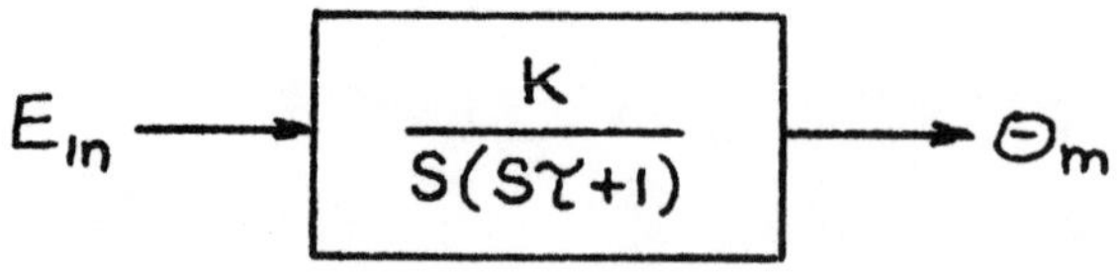

If now one considers a complete system, an error detector, and a power voltage amplifier feeding the motor:

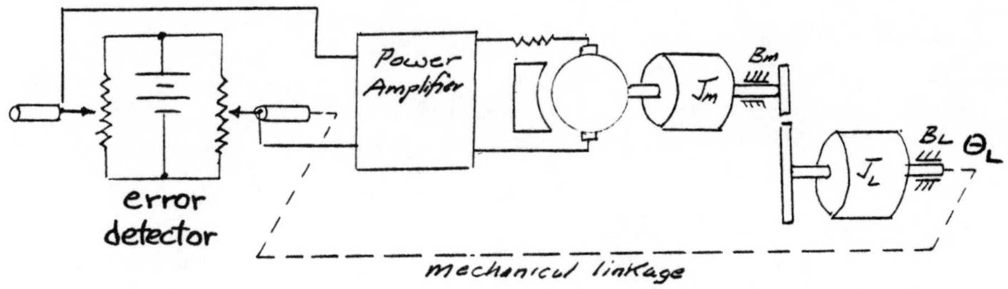

For the error detector assume potentiometers turn through 2π and a voltage source of 6.28 volts, then the block diagram becomes:

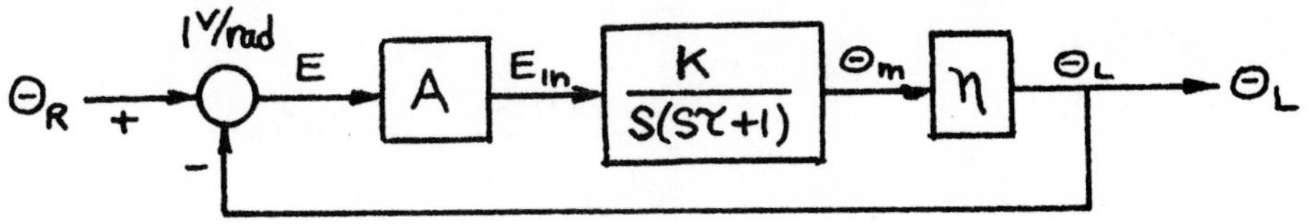

SYSTEM STABILITY

As mentioned before this can be found by several techniques, the easiest being the Routh-Hurwitz array. This involves examining the denominator of the closed loop system function :

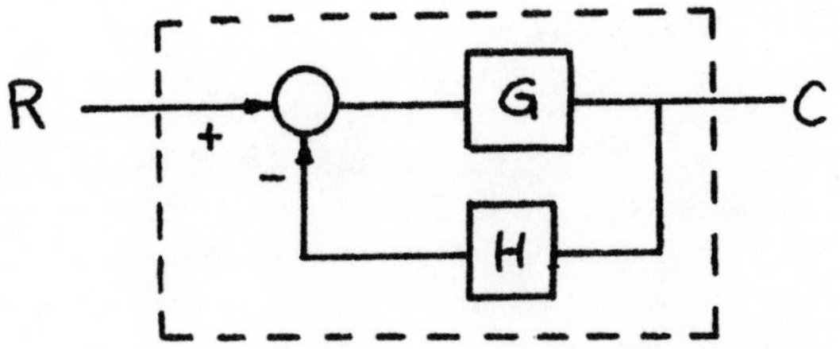

$$G_{SYST} = \frac{C}{R} = \frac{G}{1+HG} = \frac{K(S^m + a_{m-1}S^{m-1} + \cdots a_0 S^0)}{S^n + a_{n-1}S^{n-1} + \cdots a_0}$$

The characteristic equation (the denominator polynomial) may be factored by root locus techniques to determine if there are any roots in the right-half "S" plane (which, of course, gives an unstable system), however the Rough-Hurwitz method is easily implemented (rather than trying to find the factors themselves).

Consider the following simple example:

$$G = \frac{10(S+2)}{S(S+1)(S+3)}\quad,\qquad H = (S+4)$$

$$G_{SYST} = \frac{10(S+2)}{S(S+1)(S+3)+10(S+2)(S+4)} = \frac{10(S+2)}{S^3+14S^2+63S+80}$$

The characteristic polynomial: $S^3+14S^2+63S+80$

may be arranged in a Routh-Hurwitz array as follows (for details refer to any text on classical control theory):

S^3	1	63
S^2	14	80
S'	X_1	
S^0	Y_1	

$$X_1 = \frac{(14)(63)-(1)(80)}{14} = 57.3$$

$$Y_1 = \frac{(X_1)(80)-(14)(0)}{X_1} = 80$$

Since the first column is all positive the system will be stable in closed loop. (Try this problem for H being unity and G being 3 times larger - you will get an unstable system.) Of course stability may be found by frequency response techniques (i.e., Nyquist criterion, etc.); but if only the question of stability is to be answered - then use the easy R-H method.

SYSTEM PERFORMANCE - ROOT LOCUS

The behavior of the system may be determined by the location of the closed loop roots in the "S" plane and usually determined by the two most dominate ones. Root locus is nothing more than the locus of the *closed loop poles* as some value (usually the gain *K*) is varied. The rules will not be reviewed here, but a few basic concepts will be emphasized. For example:

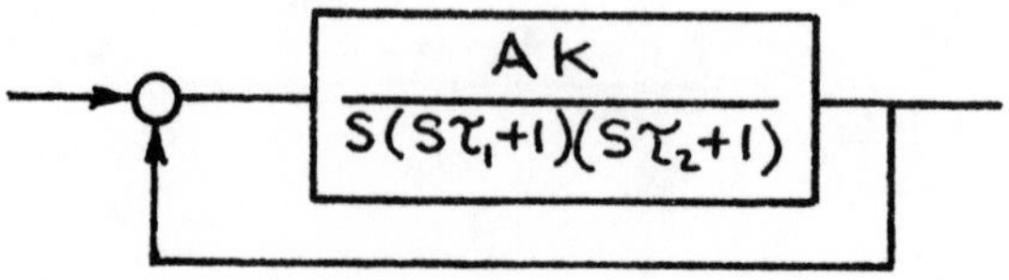

The system closed loop roots are found from solving the characteristic polynomial for various values of AK (but one must recall that the root locus gain "K_{R-L}" is not AK but is found by writing the open loop equation in terms of discrete pole values:

$$\frac{\left(\frac{AK}{\tau_1 \tau_2}\right)}{S(S+\frac{1}{\tau_1})(S+\frac{1}{\tau_2})} = \frac{K_{R-L}}{S(S+S_1)(S+S_2)}$$

Thus G_{SYST} is found as

$$G_{SYST} = \frac{G}{1+G} = \frac{K_{RL}}{S(S+S_1)(S+S_2)+K_{RL}} = \frac{K_{RL}}{S^3+a_2S^2+a_1S+a_0}$$

$$= \frac{K}{(S+\alpha)(S+\beta)(S+\gamma)}$$ where two of the roots

may be either real or complex. If one plots the root-locus, the loci eminates from the open loop poles to give the closed loop locus as:

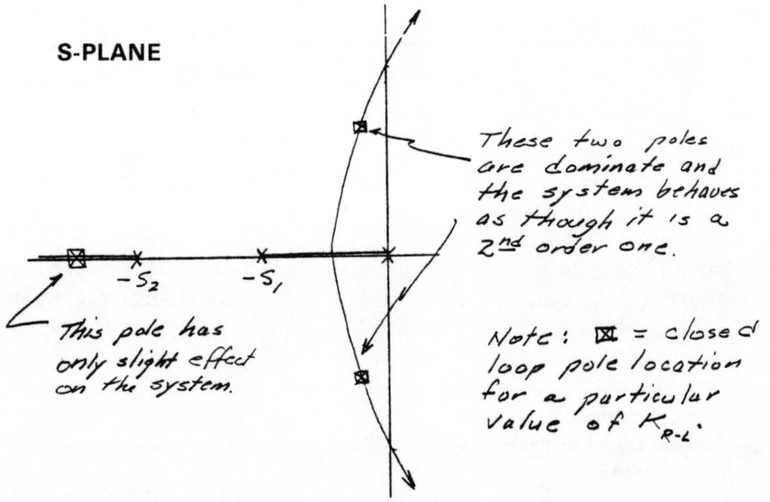
S-PLANE
These two poles are dominate and the system behaves as though it is a 2nd order one.
$-S_2$
$-S_1$
This pole has only slight effect on the system.
Note: ⊠ = closed loop pole location for a particular value of K_{R-L}.

Recall that for a second order system using standard notation of zeta (ζ), natural undamped frequency (ω_n):

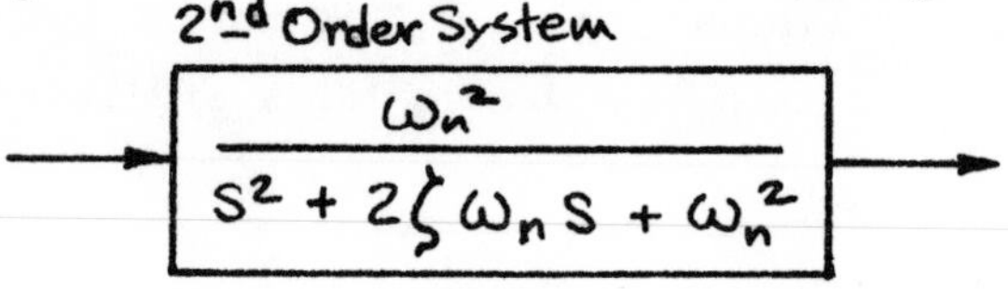

The step response is related as follows:

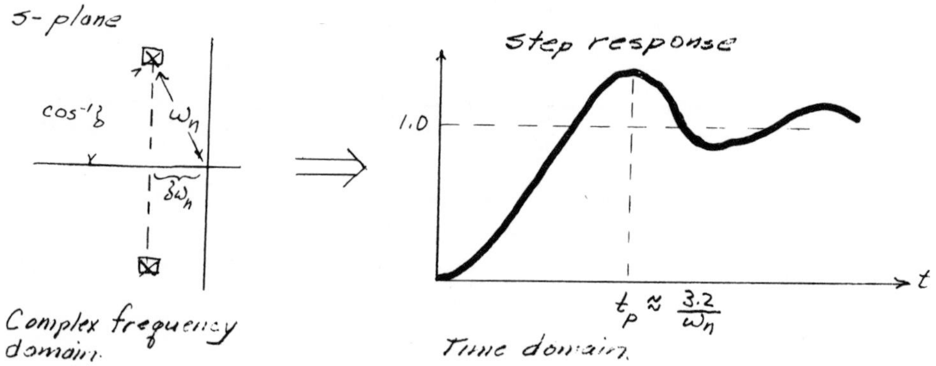

Recall (1) the longer the radius ω_n, the shorter t_p,

(2) as the angle $\cos^{-1}\zeta$ approaches 90°, the greater the overshoot,

and (3) the greater the gain, the longer ω_n, and the smaller the error. (Refer to any standardized set of curves for the second order system.)

Included in system performance is usually a statement of system final error. In short, to find the error for various standard inputs one usually makes use of the final value theorem as follows:

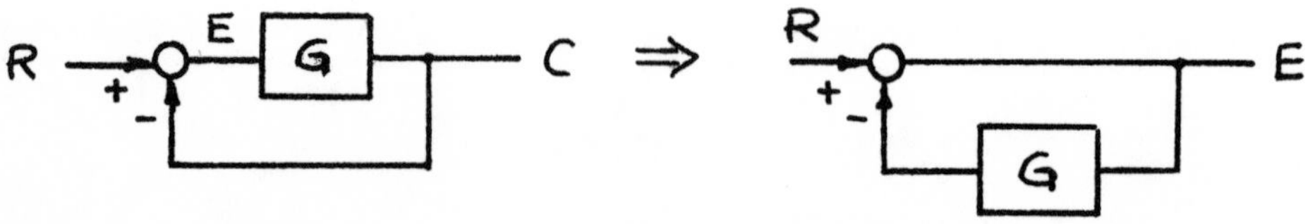

then $E = \frac{1}{1+G} R$ and using the final value theorem gives:

$$e\Big|_{t\to\infty} = \lim_{s\to 0} sE = \lim_{s\to 0} s\left[\frac{1}{1+G} R(s)\right]$$

so that if $G = \frac{K}{s(s+\alpha)}$ and the input r(t) is a unit ramp (i.e., $R = 1/s^2$) then

$$e\Big|_{t\to\infty} = \lim_{s\to 0} s\left[\frac{1/s^2}{\frac{K}{s(s+\alpha)}}\right] \to \frac{1/s}{\frac{K}{s(s+\alpha)}} = \frac{\alpha}{K}$$

(If the input had been a step, this example would have yielded zero and for an acceleration input the error would have been infinity.)

Also as part of performance specifications, frequency response is often considered. Here only brief highlights will be reviewed. Recall that if the steady state frequency response is plotted in the polar plane, the plot is the locus of the tip of the open loop phasor (assume input is a 1 volt sine wave):

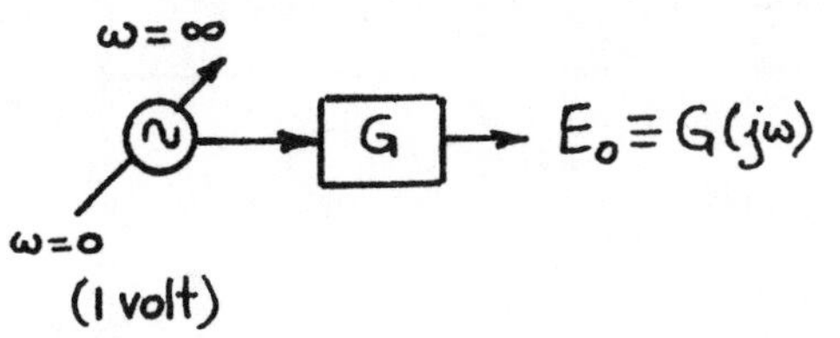

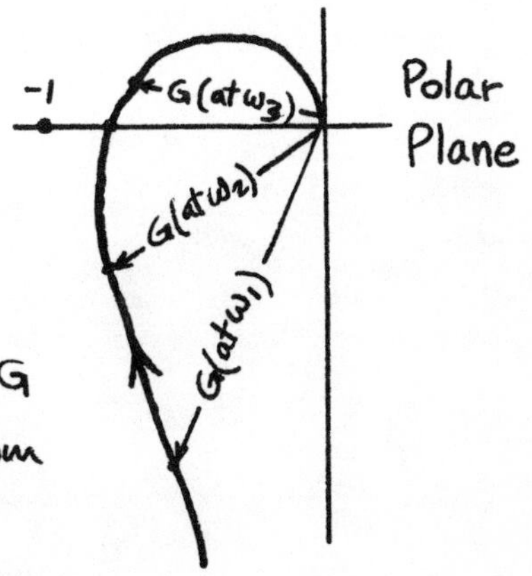

Plot of G as ω varies from 0 → ∞

Recall that (for simple systems with unity feedback) if the curve does not enclose the -1 point the closed loop system is stable (Nyquist theory); and that if the gain is increased such that the new curve would just pass through the -1 point, the amount of increase in gain necessary is called the *gain margin*. In general, the nearer the approach to the -1 point, the higher the closed loop resonant peak. *Phase margin* is defined as the amount of phase shift needed to cause the curve to pass through the -1 point (with the gain held constant); again, the smaller the phase shift needed, the higher the closed loop resonant peak. In terms of the Bode plot the gain margin is given in db and the phase margin in degrees:

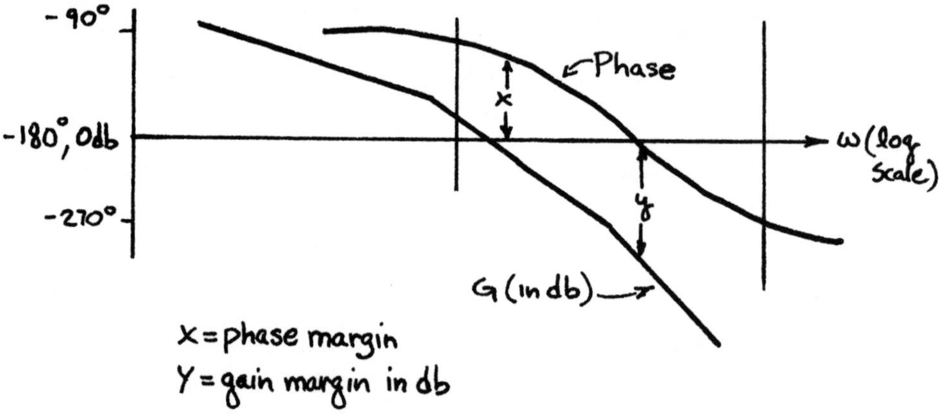

Frequently specifications are given in terms of phase margin with values being between 40° to 70° as typical.

STATE VARIABLE CONVERSION

To convert from a transfer function format to that of a state variable format, one must first recall the meaning of a state variable. Consider a standard second order system excited by a step input:

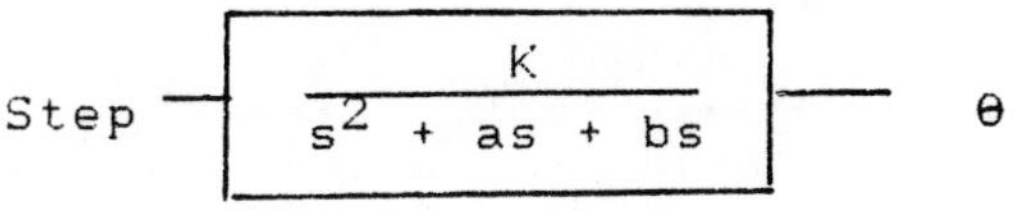

The response, θ, as a function of time for this type of function is well known and is repeated here for clarity. If one were to replot this function as the derivative - say $\dot{\theta}$, or w, vs. the function itself (i.e., w vs. θ) where time is now implicit, the plot is known as a "phase plane":

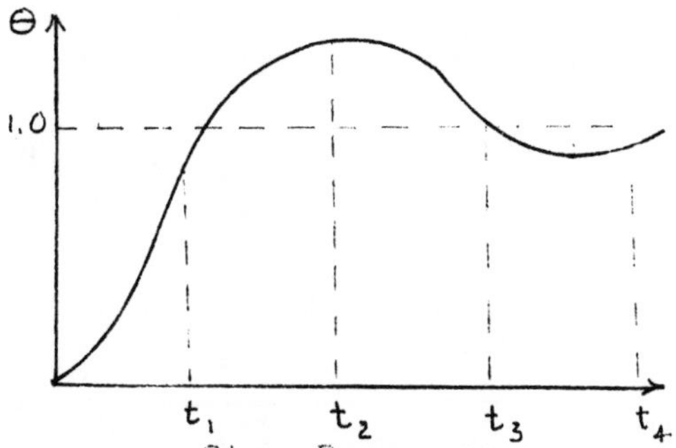

Step Response
(time explicit)

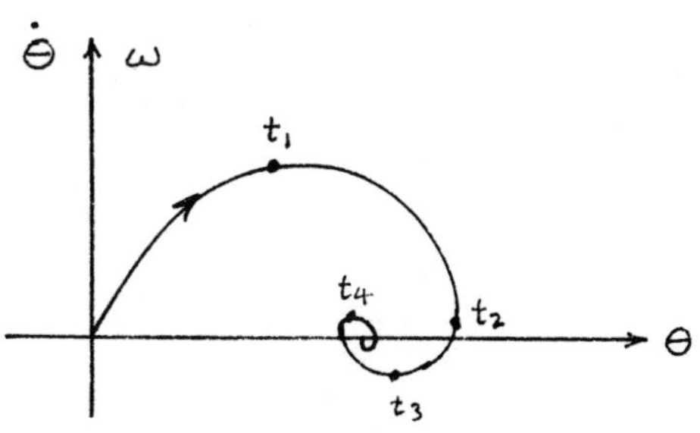

Phase Plane
(time implicit)

It is usually more convenient to redefine the variables as x_1 and x_2 (i.e., $x_1=\theta$ and $x_2=\dot{x}_1=w$), then the plot may be considered as the locus of the coordinates x_1 and x_2. The locus of the vector coordinates as a function of time is then the state variable, $\underline{x}(t)$:

For any particular time, the vector coordinates give the state of the system.

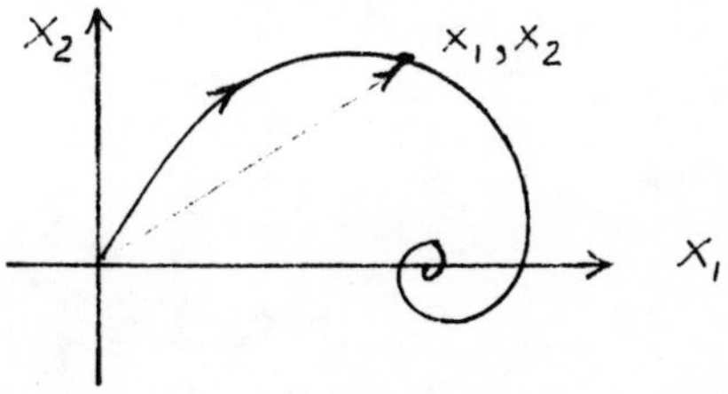

Consider the above second order system using "mixed notation" (i.e., for the Laplace $sX_1(s)$ written as $\dot{x}_1=x_2$ in the time domain - if one first considers that all initial conditions are zero), and then writes the differential equation (that the transfer function must have come from) as follows:

$$\ddot{\theta} + a\dot{\theta} + b\theta = Ke_{in} \qquad\qquad \dot{x}_1 = x_2$$

$$\text{then}$$

$$\dot{x}_2 + ax_2 + bx_1 = Ke_{in} = Ku \qquad\qquad \dot{x}_2 = -ax_1 - bx_2 + Ku$$

The mixed notation block diagram equivalent is

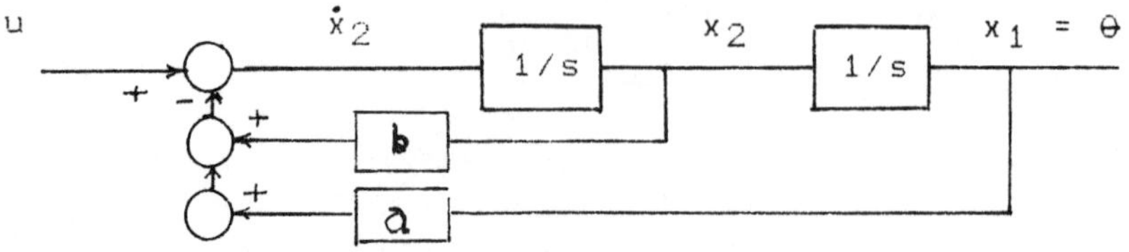

and the matrix form of the equation is given by (the underlined characters are usually written as bold faced in most texts):

$$\left.\begin{array}{l} \dot{x}_1 = x_2 \\ \dot{x}_2 = -ax_1 - bx_2 \end{array}\right\} \text{ or } \left\{ \underline{\dot{x}} = \begin{bmatrix} 0 & 1 \\ -a & -b \end{bmatrix} \underline{x} + \begin{bmatrix} 0 \\ K \end{bmatrix} u \right.$$

It follows directly from the equations that the equivalent state variable block diagram as:

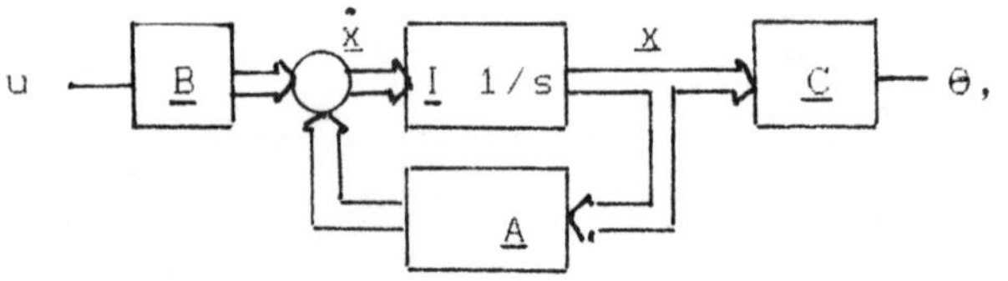

where $\underline{A} = \begin{bmatrix} 0 & 1 \\ -a & -b \end{bmatrix}$

$\underline{B} = \begin{bmatrix} 0 \\ K \end{bmatrix}$, $\underline{C} = \begin{bmatrix} 1 & 0 \end{bmatrix}$

Consider the following numerical example:

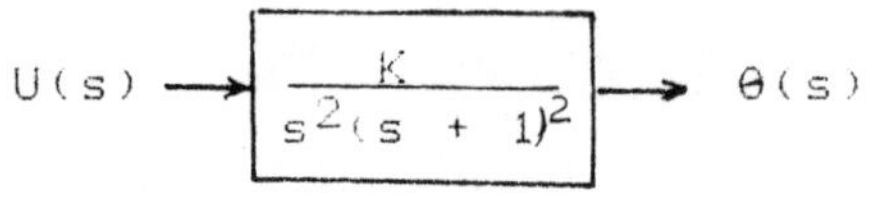

Temporarily assume all initial conditions are zero such that one may write the differential equation that the transfer function originally came from

$$\ddddot{\theta} + 2\dddot{\theta} + \ddot{\theta} = Ku$$

$$\dot{x}_4 + 2x_4 + x_3 = Ku \qquad\qquad \text{Define } x_1 = \theta,\ x_2 = \dot{x}_1 = \dot{\theta},\ \text{etc.}$$

Then, again, the mixed notation block diagram may be written as

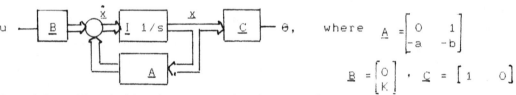

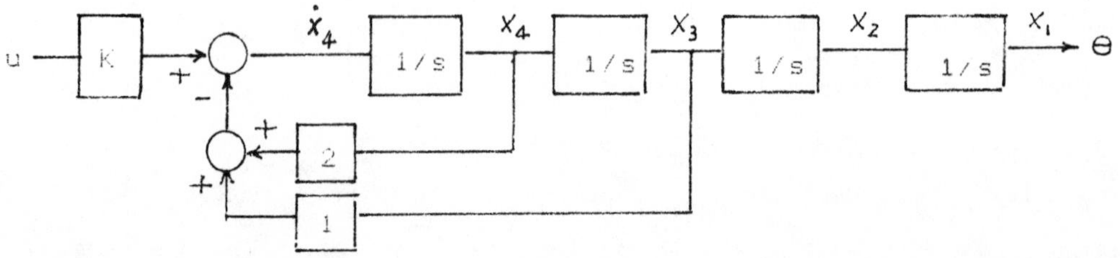

and the state variable block diagram is

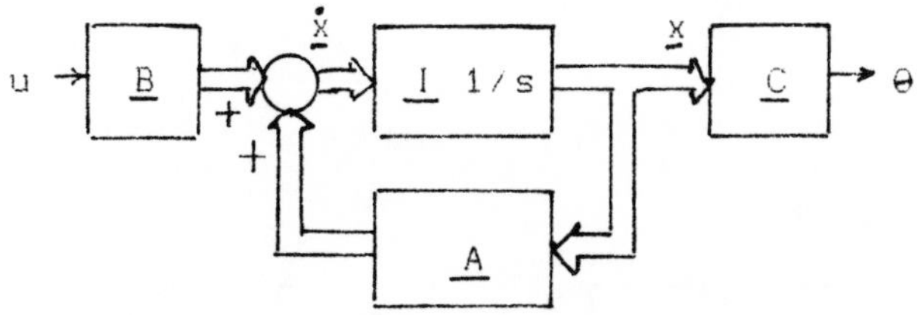

where: $\underline{A} = \begin{bmatrix} 0 & 1 & 0 & 0 \\ 0 & 0 & 1 & 0 \\ 0 & 0 & 0 & 1 \\ 0 & 0 & -1 & -2 \end{bmatrix}$

$\underline{B} = \begin{bmatrix} 0 \\ 0 \\ 0 \\ K \end{bmatrix}$, $\underline{C} = \begin{bmatrix} K & 0 & 0 & 0 \end{bmatrix}$

For a unity feedback path around the transfer function, the same form of the state variable block diagram is the same except, of course, the matrix constants change as follows (recall the rule about post- and pre- multiplying when manipulating matrices):

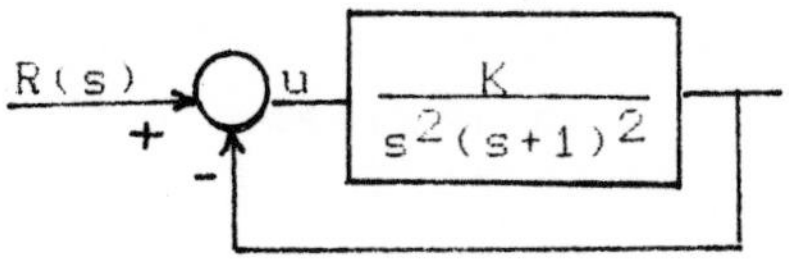

$$\dot{\underline{x}} = \underline{A}\underline{x} + \underline{B}u = \underline{A}\underline{x} + \underline{B}(r - \theta)$$
$$= \underline{A}\underline{x} + \underline{B}(r - \underline{C}\underline{x}) = (\underline{A} - \underline{B}\underline{C})\underline{x} + \underline{B}r$$

$$\dot{\underline{x}} = \underline{A}_{cl}\underline{x} + \underline{B}r.$$

The methods of solution and the operations on the equations may be found in almost any book on control systems. However, the reader must recall the meaning of solution(s); if he finds the vector $\underline{x}$ at any time, say t_1 (i.e., "freeze" the vector at t_1) then x's are merely the coordinates of the vector in the phase plane [which is easy to visualize for a phase plane of a second order system]. These coordinates are the initial conditions if one were to "unfreeze" time and start up again.

Although the probability of an in depth questions on state variables may be minimal (for one or two questions), it is recommend that one should at least review how to convert to the state variable format and then how to solve for their solutions, by either time domain or Laplace techniques. Of course there are many aspects of state variable theory one could review; in fact if the examinee specializes in control systems and intends to "tackle" all five of the questions, then one should be prepared to answer rather in depth problems in this area.

There are many simulation computer programs available for the conversion of transfer function format to the state variable presentation. Of course these computer programs also allow one to solve directly for various manipulations and the solution. Since one does not have access to these simulation programs on an examination, it is suggested that one review pencil and paper techniques on simpler problems. (You may be expected to answer questions on the subject of simulation itself for the problems in the area of computers.)

SAMPLED DATA SYSTEMS AND THE Z-TRANSFORM

For sampled systems (usually involved with computer control) one must recall that any time a signal is held or delayed, the effect is much the same as inserting the delay function, e^{-sT}, into a typical Laplace transfer function model. Thus, to gain a better feel for the behavior of a computer controlled system, consider a computer algorithm that does nothing to a signal except pass it through the computer and associated sample and hold circuit, A/D and D/A converters (for lack of a better name, use "unity thru-put"). Then the model may be thought of as

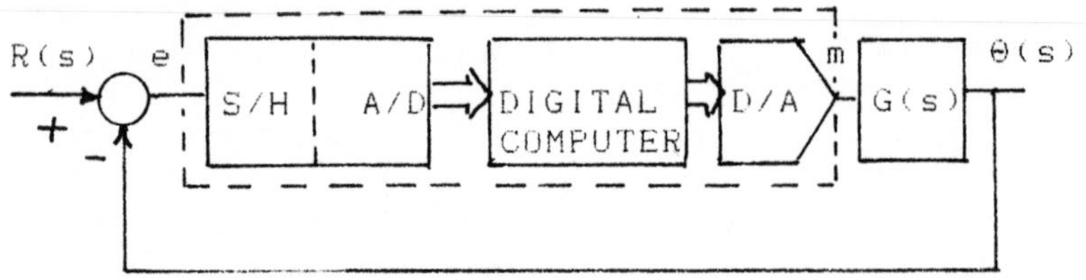

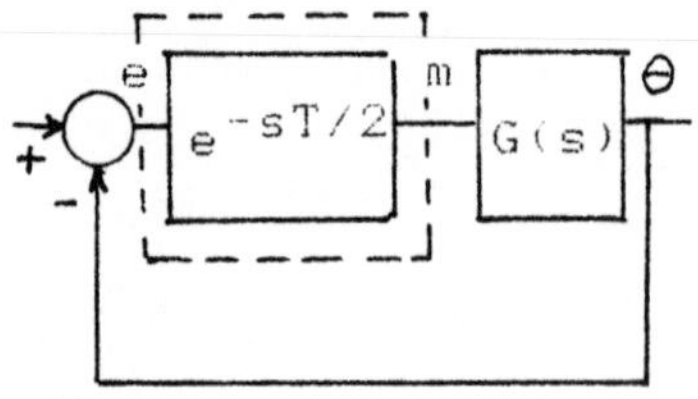

Sampled Data "unity thru-put" System

Approximate Model

An error signal wave form for a typical second order system due to a step input might have the following shape:

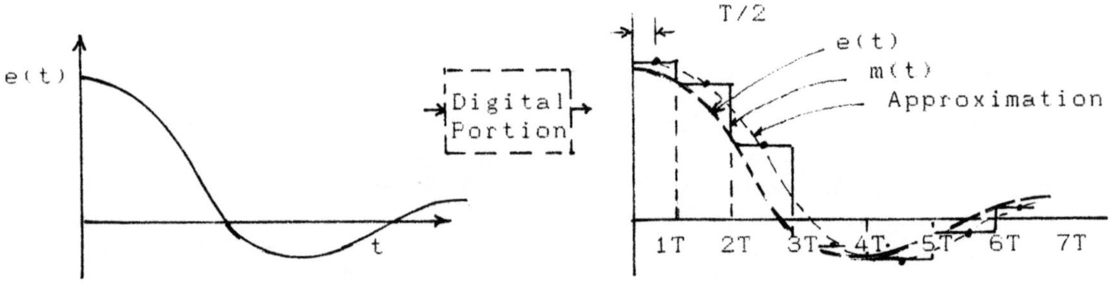

Before Delay

After S/H and Delay

The approximated signal at m(t) is the same as that at e(t) except delayed by a phase shift of 0 (or one half the time of the delayed steps). From classical theory, one may recall the effect of time delay with regard to stability by use of the Nyquist plot.

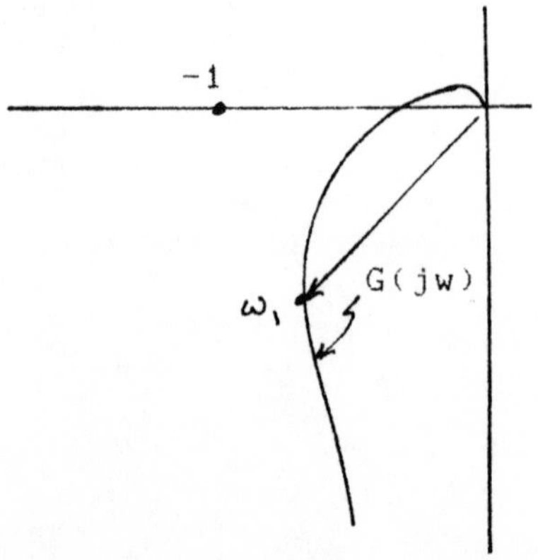

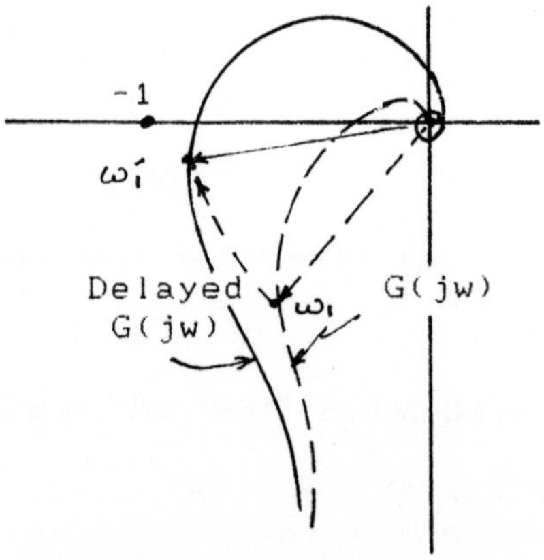

Nyquist Plot (T=0)

Nyquist Plot (T=finite)

Obviously, the system tends towards instability for greater delay and may need compensation. The Laplace delay function, e^{-sT}, is very similar to the effect of z^{-1} when using z-transforms. Thus a short review of z-transforms is in order.

Recall that a sampling switch is assumed to close only for an instant of time; then, whatever signal is sampled, the signal is held until the next sample time. If the Laplace of the unsampled signal is E(s), then the delayed sampled signal may be given by

$$\mathcal{L}[f*(t)] = \sum_{k=0}^{\infty} f(kT)e^{-kTs} = \sum_{k=0}^{\infty} f(kT)z^{-kk}, \quad \text{where } z = e^{sT}$$

Then by direct substitution of a particular function, say a STEP, into the equation would yield the z-transform for that function.

$$\mathcal{Z}[u(t)] = \sum_{k=0}^{\infty} u(kT)z^{-k} = \sum_{k=0}^{\infty} z^{-k} = \frac{z}{z-1}$$

where z^{-k} may be represented by its geometric infinite series as a closed function of

$$z^{-k} = 1 + z^{-1} + z^{-2} + z^{-3} + \ldots = \frac{z}{z-1}$$

The z-transform for several functions may be shown to be

	f(t)	F(s)	F(z)
Step	$u(t)$	$1/s$	$z/(z-1)$
Ramp	$tu(t)$	$1/s^2$	$Tz/(z-1)^2$
Exp	e^{-at}	$1/(s+a)$	$z/(z-e^{-aT})$
Sine	$(\sin wt)u(t)$	$w/(s^2+s^2)$	$z \sin wT/(z^2-2z \cos wT +1)$

The sample and hold (with the time delay) has, as its Laplace representation, the effect of

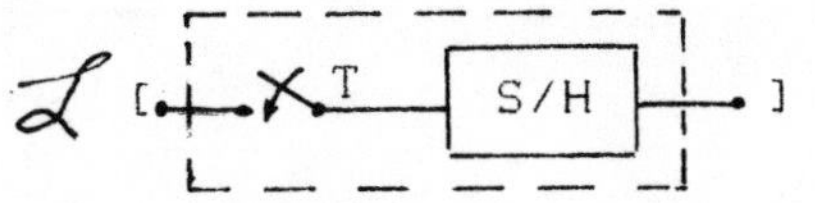

$$= (1 - e^{-sT})/s$$

And, the method of analysis, if the transfer function (in "s") is known, is to make the whole system as a Laplace transform, <u>then convert to the z-transform as a whole</u>. A simple example will illustrate the method:

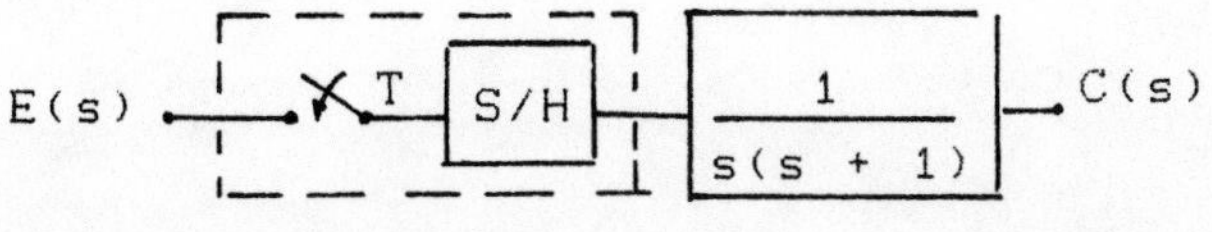

$$G_x(s) = \frac{1 - e^{-sT}}{s^2(s+1)^2} = C/E$$

Then, to convert to the z-transform, merely expand the total Laplace transform by the partial fraction method, and convert directly to the desired result (recall, $e^{-sT} = z^{-1}$):

$$G_x(s) = (1 - e^{-st})[\frac{1}{s^2} - \frac{1}{s} + \frac{1}{s+1}] \qquad (1-z^{-1})[\frac{Tz}{(z-1)^2} - \frac{z}{z-1} + \frac{z}{z-e^{-T}}]$$

One may find the response to a unit impulse input exactly like the method for the Laplace transform technique (i.e., multiplying the functions - in "z" - together)

$$C(z) = [G(z)](1) \qquad \text{since} \qquad [\text{Impulse}] = 1$$

Or, for a numerical value, assume the sample period is known to be one second, then simplifying

$$C(z) = [\frac{(ze^{-T} - z + Tz) + (1 - e^{-T} - Te^{-T})}{(z-1)(z-e^{-T})}]\Bigg|_{T=1} = \frac{0.368z + 0.264}{z^2 - 1.368z + 0.368}.$$

Finally the answer for the step response may be found by several different methods; perhaps the easiest way is merely to divide the numerator by denominator

$$z^2 - 1.368z + 0.368 \overline{\big)\, 0.368z + 0.264} \quad \text{quotient: } 0.368z^{-1} + 0.767z^{-2} + 0.914z^{-3} + 0.968z^{-4} + \ldots$$

The response (in "z") is directly converted to a difference equation (in "k") and gives the result at discrete times (with no information about points in between the sample periods, kT):

$$c(kT) = 0.368(k=1) + 0.762(k=2) + $$
$$+ 0.914(k=3) + 0.968(k=4) + \ldots$$

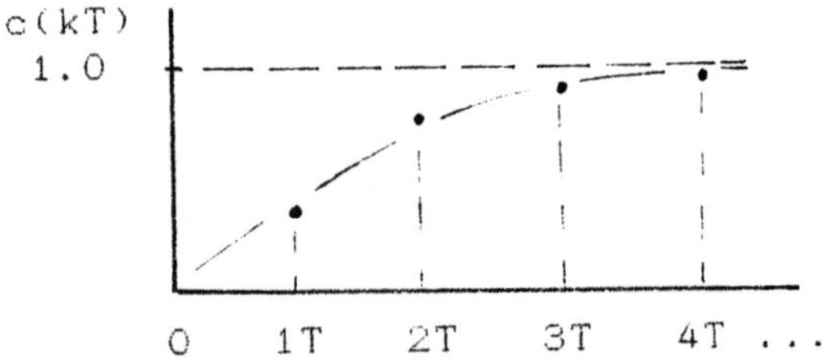

For a closed loop system, the method is the same

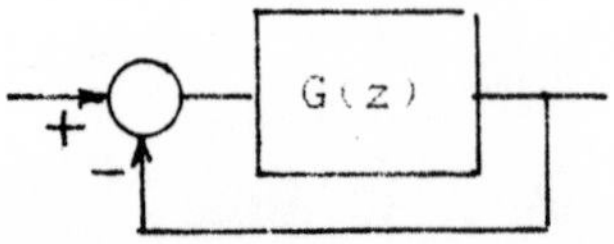

$$\Longrightarrow \qquad G_{cl}(z) = \frac{G(z)}{1 + G(z)}$$

To find whether the system is stable, recall that for a system in "s", if there are no poles in the right half plane, the system is stable. That is, the "jw" axis is the dividing line. Consider the equation e^{-sT}, where $sT = aT + jwT$; jwT represent the angle, and e^{aT} represents the magnitude of z. Therefore the dividing line is when alpha equals zero, or the absolute value of z is unity. Thus, the jw axis is the s-plane is equivalent to the unit circle in the z-plane:

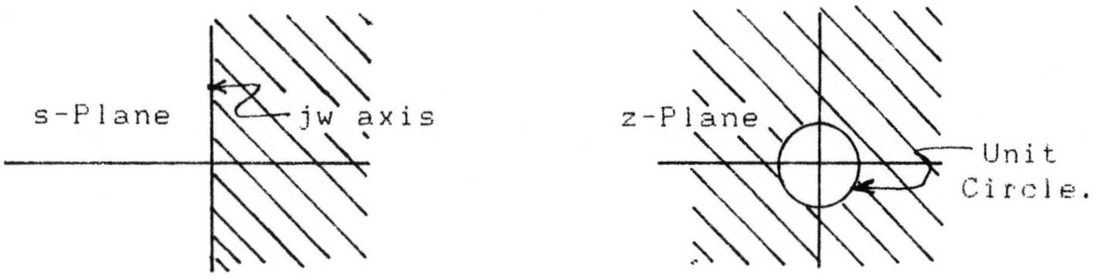

Stability Limits

The stability of the system may easily be determined by calculating the poles of the characteristic polynomial (CP) of the function (in z), and determining if they are <u>within</u> the unit circle (again assume T= one second):

$$1 + G(z) = 1 + \frac{0.368 + 0.264}{z^2 - 1.368z + 0.368} = \frac{z^2 - z + 0.632}{z^2 - 1.368z + 0.368}$$

$$CP = z^2 - z + 0.632 = (z - 0.5 + j0.618)(z - 0.5 + j0.618)$$

For the closed loop system (assuming the sample period is unity), the characteristic polynomial has roots within the circle and is therefore stable. One should note that stability (for a linear system) does not depend upon the input.

Question: Repeat the above sample problem (in closed loop) but let the input be a unit step. Also, find whether the system is stable.

Solution: The same method as above except the system is excited by the step, $z/(z-1)$, rather than unity. The output equation (in z) becomes (again for T=unity)

$$c(kT) = 0.368(k=1) + 1.0(k=2) + 1.4(k=3) + 1.4(k=4) + 1.15(k=5) + \ldots$$

For stability, the answer is the same as before since stability is independent of the excitation.

Again, for those whose who specialize in the area of control systems and attempts to work all five problems, they should be prepared to answer more "in depth" material than is presented here.

CONTROL SYSTEMS 1

This is a 10-question multiple choice problem. It may be longer and more detailed than might be expected on the actual exam.

The positional movement of a robot arm has a theoretical open loop transfer function of:

$$G(s) = C(s)/E(s) = K(s+4)/(s(s+1)(s+2))$$

QUESTION 1 If an optical sensor is used to monitor the arm position such that a closed loop system is achieved, what is the closest gain, K, that would just cause system instability (assume the electronic optical sensor unit is equivalent to unity feedback)?

(a) Zero
(b) Infinity
(c) 2.0
(d) 4.0
(e) 6.0

SOLUTION Use Routh-Hurwitz method:

$$G_{sys} = \frac{G}{1+G} = \frac{K(s+4)}{s^3 + 3s^2 + (2+K)s + 4K}$$

s^3	1	$(2+K)$
s^2	3	$4K$
s^1	x_1	
s^0		

$$x_1 = \frac{3(2+K) - 4K}{3} > 0$$

$$\therefore 6 - K > 0$$

$$K < 6$$

Answer is (e).

QUESTION 2

In open loop the arm transfer function has a frequency response such that the phase shift will be -180° at a frequency nearest to what value (in rad/sec)?

(a) Zero
(b) Infinity
(c) Unable to determine since K has not been specified.
(d) 3.0
(e) 5.0

SOLUTION Use the Bode phase approximation method; sketch on semi-log paper, or use the direct calculation method with two or three "guessed" trial frequencies:

$$G = \frac{K(s+4)}{s(s+1)(s+2)} = \frac{2K(0.25s+1)}{s(s+1)(0.5s+1)}$$

$s \rightarrow j\omega$

$$\angle\phi = \tan^{-1} 0.25\omega - 90° - \tan^{-1}\omega - \tan^{-1} 0.5\omega \Rightarrow -180°$$

Try $\omega = 2$:

$$\angle\phi = 14° - 90° - 63.4° - 26.6° = -166° \neq 180°$$

Try $\omega = 3$:

$$\angle\phi = 36.9° - 90° - 71.6° - 56.3° = -181° \approx -180$$

Answer is (d).

QUESTION 3

In closed loop (with the unity feedback sensor connected) the percentage overshoot of the output is to be near 16% for a step input; what would be the approximate damping ratio, zeta, (if the system may be approximated by a second ordered one).

(a) Over damped (nonexistent).
(b) Very lightly under damped.
(c) 0.5
(d) 0.7
(e) 1.0

SOLUTION Since the system is approximated by a second-ordered one, the normalized, standardized curves give a family of plots that relate zeta to percent overshoot. Or, the damping ratio may be calculated directly:

$$\%\,O.S. = 100\,e^{-\zeta\pi/\sqrt{1-\zeta^2}}, \quad 0.16 = e^{-\zeta\pi/\sqrt{1-\zeta^2}}$$

$$\ln 0.16 = -1.833 = \zeta\pi/\sqrt{1-\zeta^2}, \quad (1.833)^2 = \frac{\zeta^2\pi^2}{1-\zeta^2}, \quad \zeta \doteq 0.5$$

Answer is (c).

QUESTION 4

In closed loop (with the unity feedback sensor connected) and with an increased gain setting, it is found that the system will just break into oscillation. What is the closest natural frequency (in rad/sec) of this oscillation?

(a) Eventually reaches infinity
(b) Will never break into oscillation
(c) 1.0
(d) 3.0
(e) 6.0

SOLUTION From the root locus plot, find the frequency which the curve just passes through the imaginary axis; or, the value may be calculated directly from the closed loop characteristic polynomial (CP) equation:

$$C.P. = s^3 + 3s^2 + (2+K)s + 4K = 0, \quad \text{for } s = 0 + j\omega \text{ (on axis)}$$

$$= -j\omega\omega^2 - 3\omega^2 + j\omega(2+K) + 4K = 0$$

$$= \underbrace{(-3\omega^2 + 4K)}_{0} + \underbrace{j\omega(-\omega^2 + 2 + K)}_{0} = 0$$

$$\therefore \omega^2 = 8, \quad \omega = \sqrt{8} \approx 3$$

Answer is (d).

QUESTION 5

For the system in closed loop (through the optical sensor, for unity feedback) the gain has been set such that the damping ratio (zeta) of 0.5, what percent overshoot may be expected for a step input? (Assume the system is approximated by a second-ordered one.)

(a) 16%
(b) 25%
(c) 37%
(d) 50%
(e) Unable to determine since K is not given.

SOLUTION This is a "standard" graphical relationship that that is given in almost any text on control system that is plotted for an ideal second-order curve. Or it may be calculated from:

$$\% \text{ O.S.} = 100 e^{-\delta\pi/\sqrt{1-\delta^2}} = 100 e^{-0.5\pi/\sqrt{1-\delta^2}}$$
$$= 16.3 \approx 16$$

Answer is (a).

QUESTION 6

For the system in closed loop (unity feedback) to have adamped natural frequency of 1.0 rad/sec, what is the closest gain setting for K?

(a) 6.0
(b) 4.0
(c) 2.0
(d) 0.5
(e) Unable to determine since zeta is not given.

SOLUTION Carefully sketch the root locus in the region near 1.0 rad/sec (at the intersection of the crude root locus and a straight line passing through the imaginary axis at 1.0). Once the correct locus is determined, K is the product of the line lengths from the open loop poles divided the line length of the open loop zero:

$$K = \frac{l_1 l_2 l_3}{l_0}$$
$$= \frac{1.1 \times 1.2 \times 1.8}{3.8} = 0.6$$

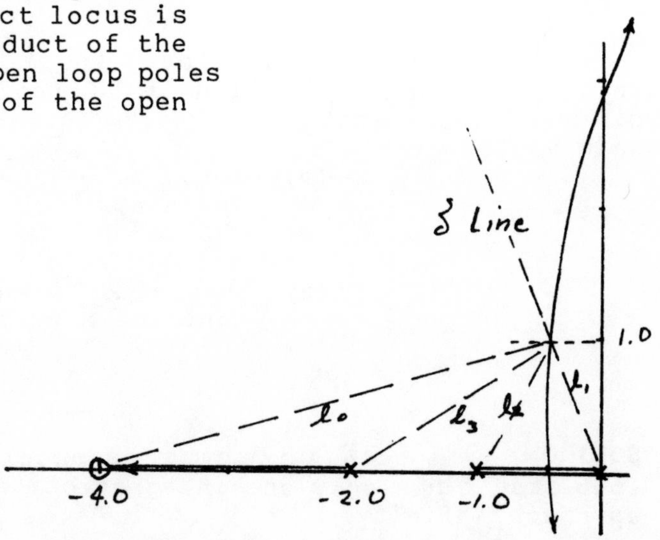

Answer is (d).

QUESTION 7

The arm (in closed loop) is to follow a moving object with a slew rate (ramp) of 2.0 rad/sec. What is the closest steady state error of the arm and the object for a system gain K=4.0? The arm (in closed loop) is to follow a moving object with a slew rate (ramp) of 2.0 rad/sec. What is the closest steady state error of the arm and the object for a system gain K=4.0?

(a) Zero
(b) Infinity (afterinfinite time).
(c) 0.25
(d) 4.0
(e) 10.0

SOLUTION Since this is a type I system (i.e., one integration in the loop equation, the error will be finite and is found from the final value theorem:

$$E = \frac{1}{1+G} R$$

$$e(t)\Big|_{t \to \infty} = \lim_{s \to 0} s \left[\frac{1}{1 + \frac{K(s+4)}{s(s+1)(s+2)}} \left(\frac{2}{s^2}\right) \right] \to \frac{2}{\frac{4(4)}{(1)(2)}} = 0.25$$

Answer is (c).

QUESTION 8

The arm (in closed loop) is to follow a moving object with an acceleration of 1.0 rad/sec/sec. What is the closest steady state error of the arm and the object for a system gain K=1.0?

(a) Zero
(b) Infinity
(c) 1.0
(d) 4.0
(e) Unable to calculate since transfer function contains a zero in the numerator.

SOLUTION Since the input is an acceleration ($R(s)=1/s^3$), the loop itself has only one integration, the error will have to approach infinity.

Answer is (b).

QUESTION 9

For the closed loop system and for a particular gain that yeilds a damped natural frequency of 1.0 rad/sec, the system is too slow, but has an acceptable damping ratio. A faster system with a higher gain and damped natural frequency of 1.5 rad/sec meets time specifications but has too much overshoot. To remedy the situation, a phase lead compensator (a zero and pole combination) is to be used in a feedforward configuration. The compensator zero is to have a value of 1.0 (i.e., -1 in the s-plane), what is the approximate value of the compensator pole such that the damping ratio, zeta, is near that of original system (i.e., when the frequency was 1.0, yet the frequency is 1.5)?

(a) Unable to calculate as K is unknown.
(b) 1.0
(c) 2.5
(d) 4.0
(e) 25

SOLUTION This is somewhat lengthy to solve (i.e., time wise, it may pay to skip this question). One method is to use the original root locus to locate the original uncompensated zeta for a damped frequency of 1.0 (i.e., a horizontal line passing through the imaginary axis at 1.0 and the root locus); using this zeta line, locate the desired new locus location on the zeta line and a horizontal line passing through the imaginary axis at 1.5. This new locus location will be incorrect by a certain angle (i.e., all correct locus locations should add to $+/-n180^{o}$); find the difference between this certain angle and that of -180° (this difference will be approximately 35°). Then, by trial and error, locate the compensator pole on the real axis such that the new locus angles will sum to -180° (a location of -2.6 for this pole will satisfy the angle relationship).

Answer is (c)

QUESTION 10

The open loop function has a state variable representation of:

$$\underline{x} = \underline{A}\underline{x} + \underline{B}e, \quad c = \underline{C}\underline{x} = y.$$

Note: Do not confuse the lower case "c" (i.e., the output of the transfer function) with the upper case "$\underline{C}$" (i.e., the connecting matrix).

And, if it is known that $\underline{A} = \begin{matrix} 0 & 1 & 0 \\ 0 & 0 & 1 \\ 0 & -2 & -3 \end{matrix}$, then the matrix $\underline{C}$ that is representative of the original transfer function is closest to which of the following row matrix?

(a) [4K K 0]
(b) [K 4K 0]
(c) [1 2 3]
(d) [0 -2 -3]
(e) None of above since order of $\underline{A}$ matrix is incorrect.

SOLUTION General the $\underline{A}$ matrix may take several forms, however here it is given, the $\underline{C}$ matrix relating the output to the state variables may be found directly as follows:

$$G = \frac{Ks + 4K}{s^3 + 3s^2 + 2s} = \frac{Ks^{-2} + 4Ks^{-3}}{1 + (3s^{-1} + 2s^{-2} + 0s^{-3})}$$

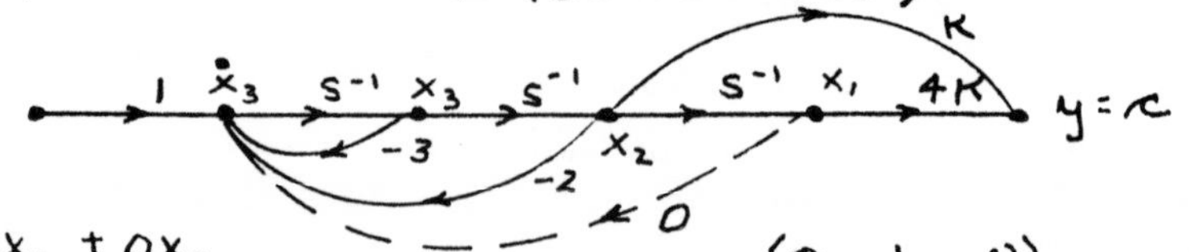

$$\dot{x}_1 = 0x_1 + 1x_2 + 0x_3$$

$$\dot{x}_2 = 0x_1 + 0x_2 + 1x_3$$

$$\dot{x}_3 = 0x_1 - 2x_2 - 3x_3 + 1e$$

$$y = c = 4K + Kx_2 + 0x_3$$

$$\dot{\underline{x}} = \begin{bmatrix} 0 & 1 & 0 \\ 0 & 0 & 1 \\ 0 & -2 & -3 \end{bmatrix} \underline{x} + \begin{bmatrix} 0 \\ 0 \\ 1 \end{bmatrix} e$$

$$y = [4K \quad K \quad 0]\,\underline{x} = c$$

Answer is (a)

CONTROL SYSTEMS 2

Determine whether the following system is stable and predict the closed loop pole location for the system for $K = 4$.
Also, find the system error.

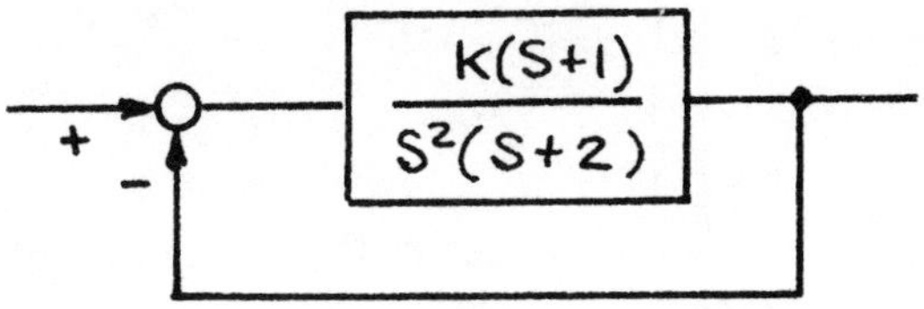

SOLUTION

For system stability use Routh-Hurwitz method:

$$G_{SYST} = \frac{G}{1+G} = \frac{K(S+1)}{S^2(S+2)+K(S+1)} = \frac{K(S+1)}{S^3+2S^2+KS+K}$$

Characteristic Polynominal: S^3+2S^2+KS+K

Routhian Array

S^3	1	K
S^2	2	K
S^1	X	
S^0	Y	

where $X = \frac{(2)(K)-(K)}{2} = \frac{K}{2}$

and $Y = \frac{(K/2)(K)-0}{(K/2)} = K$

Therefore the first column is positive for all positive values of K and the system is stable.

The root locus is sketched by setting $G = -1$ (that is, $|G|=1$, $\Phi_G = \pm n180°$ with n being any odd integer), then using the basic rules of root locus, one obtains the following sketch :

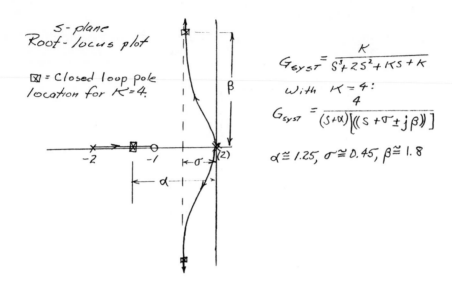

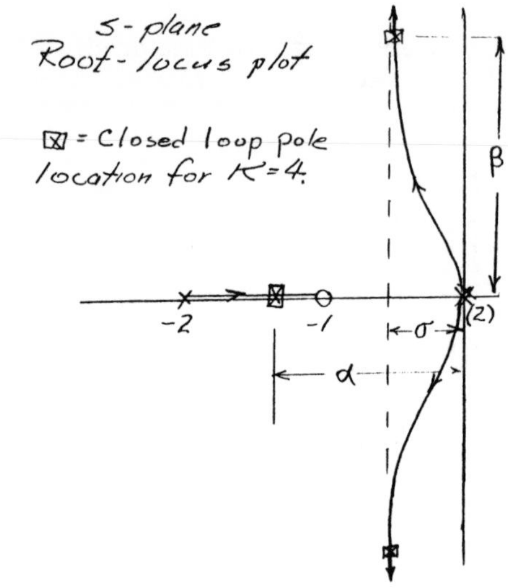

$$G_{syst} = \frac{K}{s^3 + 2s^2 + Ks + K}$$

with $K = 4$:

$$G_{syst} = \frac{4}{(s+\alpha)[(s+\sigma \pm j\beta)]}$$

$\alpha \cong 1.25$, $\sigma \cong 0.45$, $\beta \cong 1.8$

The system is stable for all values of positives K's.

The system error is zero for both a step and ramp input, but is finite for an acceleration input (a/s^3):

$$e\Big|_{t\to\infty} = \lim_{s\to 0} s\left[\frac{R}{1+G}\right] = \lim_{s\to 0} s\left[\frac{(a/s^3)}{1 + \dfrac{4(s+1)}{s^2(s+2)}}\right]$$

$$= \lim_{s\to 0} \frac{a}{s^2 + \dfrac{4(s+1)}{s+2}} = \frac{a}{(4/2)} = 0.5\,a$$

CONTROL SYSTEMS 3

The open loop transfer function for a control system is approximated by:

$$G(s) = \frac{C(s)}{E(s)} = K \frac{s - 3}{(s + 0.5)(s + 7)}$$

It is desired to make the output signal (C) correspond as nearly as possible to some input signal, (R), in steady state, at the same time keeping the system stable.

REQUIRED:

Wt.		
3	(a)	Sketch a block diagram for a feedback control system to accomplish the given objective. Carefully label the summation polarity of all signals coming into the feedback junction summing point. (Note that the given transfer function has peculiar properties.)
7	(b)	Select a value of K which assures system stability and at the same time brings the ratio $\frac{C}{R}$ in steady-state as close to +1.0 as possible. (Note that the properties of G(s) are such that it is advisable to make a very careful check on the requirements for closed-loop system stability.)

SOLUTION

a) The block diagram could be given as follows (with either positive or negative feedback):

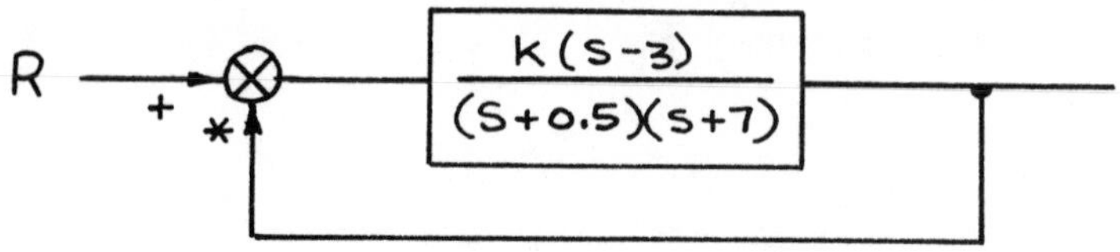

*For negative (-) feedback, the system function is:

$$G_{SYST} = \frac{G}{1+G} = \frac{K(S-3)}{(S+0.5)(S+7) + K(S-3)}$$

$$= \frac{K(S-3)}{S^2 + (7.5+K)S + (3.5-3K)}$$

The denominator (which determines the "character" of the response) must not have any negative factors (indicating closed loop poles in the right half plane -- or an unstable system). A simple test for stability is the Routh criterion; or in this case, (for a simple second-order system) all of the coefficients of the denominator polynomial must be positive, therefore:

$$3.5 - 3K > 0, \quad \therefore \; K < \frac{3.5}{3} = 1.17$$

*For positive (+) feedback, the system function is:

$$G_{SYST} = \frac{G}{1-G} = \frac{K(S-3)}{(S+0.5)(S+7) - K(S-3)}$$

$$= \frac{K(S-3)}{S^2 + (7.5-K)S + (3.5+3K)}$$

Here, for stability, again the denominator polynomial must be positive, thus:

$$K < 7.5$$

Of course the root locus method of analysis may also be used:

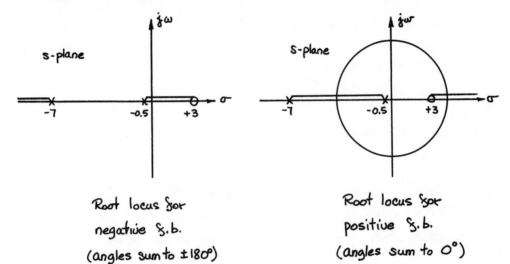

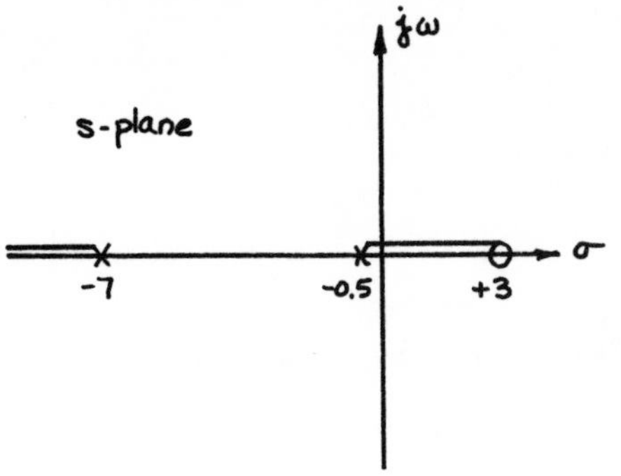

Root locus for negative f.b. (angles sum to ±180°)

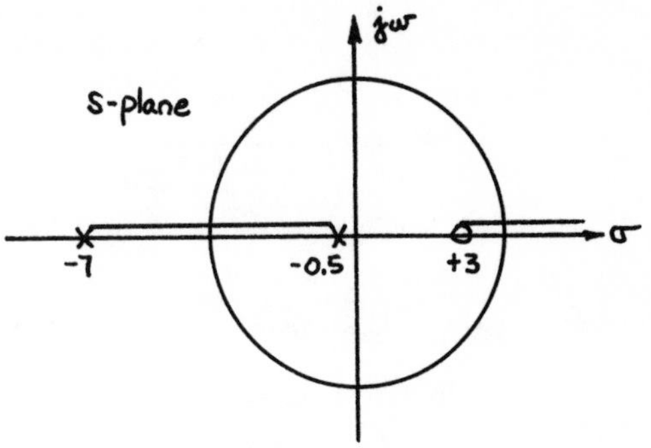

Root locus for positive f.b. (angles sum to 0°)

b) For $\frac{C}{R}$ to be as close to unity as possible, then it is only necessary to minimize $E = R - C$ (since it is a unity feedback system):

For negative feedback:

$$E = \frac{R}{1 + G}$$

Now assume a step input $\left(R = \frac{r_o}{s}\right)$ and that one is interested in the error after a long period of time -- such that the final value theorem may be applied:

$$e\Big|_{t\to\infty} = \lim_{s\to 0} sE = \lim_{s\to 0} s\left[\frac{r_o/s}{1+\frac{K(s-3)}{(s+0.5)(s+7)}}\right]$$

$$= \frac{r_o}{1+\frac{K(-3)}{(0.5)(7)}} = \frac{r_o}{1-0.85K}$$

To minimize this, $(1-0.85K)$ should be as large as possible. However, since the maximum value of K that can be used (for system stability) makes $0.85K=1$, therefore it is obvious that for the smallest error for negative feedback, $K \to 0$.

Consider positive feedback:

$$E = \frac{R}{1-G}$$

then

$$e\Big|_{t\to\infty} = \lim_{s\to 0} s\left[\frac{r_o/s}{1-\frac{K(s-3)}{(s+0.5)(s+7)}}\right]$$

$$= \frac{r_o}{1+\frac{3K}{3.5}}$$

and here K can take on values up to 7.5 for a

stable system, therefore the minimum value for "e" will be when K just equals 7.5:

$$e = \frac{r_o}{1+6.43} = \frac{r_o}{7.43}$$

Thus, for this peculiar configuration, positive feed back gives the smallest system error and the value of K may vary from 0 to 7.5.

CONTROL SYSTEMS 4

CONTROL: SYSTEM STABILITY

Consider the following simplified diagram for a control system:

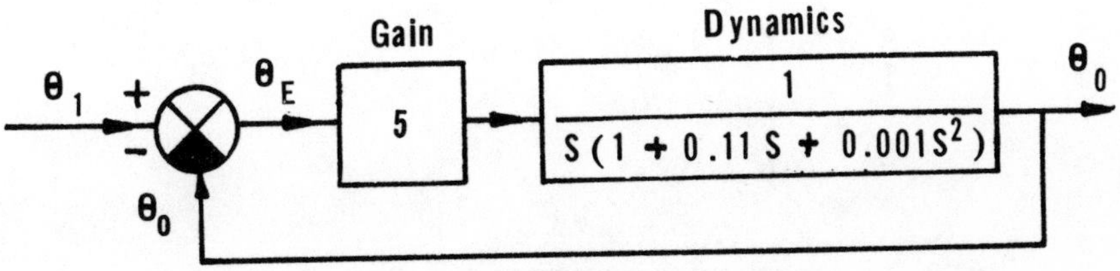

(1) Wt. 6 Determine whether the system is stable by any standard criterion.

(2) Wt. 4 What range of real values of gain will enable the system to be stable?

SOLUTION

There are several methods of determining system stability, two of which are:

1. Routh's criterion*
2. Root locus**

All of these methods are based upon solving the equivalent closed loop "characteristic equation" for positive roots, i.e., all solutions that would indicate whether the resulting exponential terms were expanding with time.

Routh Criterion: Solve for the closed loop transfer function:

$$KG_{CL} = \frac{KG}{1+KG} = \frac{5}{0.001S^3 + 0.11S^2 + S + 5}$$

and the characteristic equation (denominator) can be formed into a triangular array as follows:

Characteristic eqn.:

$$A_n S^n + A_{n-1} S^{n-1} + A_{n-2} S^{n-2} + ---- A_o$$

Then:

S^n	A_n	A_{n-2}	A_{n-4}	- - -
S^{n-1}	A_{n-1}	A_{n-3}	A_{n-5}	- - -
S^{n-2}	X_a	X_b	X_c	- - -
⋮	⋮			
S^o	Z_a			

Where $X_a = \dfrac{A_{n-1}A_{n-2} - A_n A_{n-3}}{A_{n-1}}$ etc.

All coefficients of the first coefficient column must be of the same sign for stability.

---

* Thaler and Brown, "Servo-mechanism Analysis", McGraw-Hill, p 151.

** Ibid, pp 304 - 313.

Thus:

S^3	0.001	1
S^2	0.11	5
S^1	0.955	
S^o	5	

where

$$X_1 = \frac{(.11)(1) - (0.001)(5)}{0.11} = 0.955$$

$$Y_1 = \frac{(0.955)(5) - 0}{.955} = 5$$

Therefore system is stable for the given value of gain.

(2) To determine the range of values of gain for the system to be stable, merely let the "5" in the Routh-Hurwitz array be a variable K and solve:

S^3	0.001	1
S^2	0.11	K
S^1	X_1	

$$X_1 = \frac{(.11)(1) - (0.001)\,K}{0.11}$$

and since the requirement for X_1 is that it be positive for stability:

$$0.11 > 0.001\ K$$

$$K < 110 \text{ for stability}$$

Solution based upon root-locus method:

$$KG = \frac{K}{S(1 + 0.11S + 0.001S^2)} = \frac{K/1000}{S(1000 + 110S + S^2)}$$

$$= \left(\frac{K}{1000}\right)\left[\frac{1}{S(S + 10)(S + 100)}\right]$$

The root-locus plot will then be:

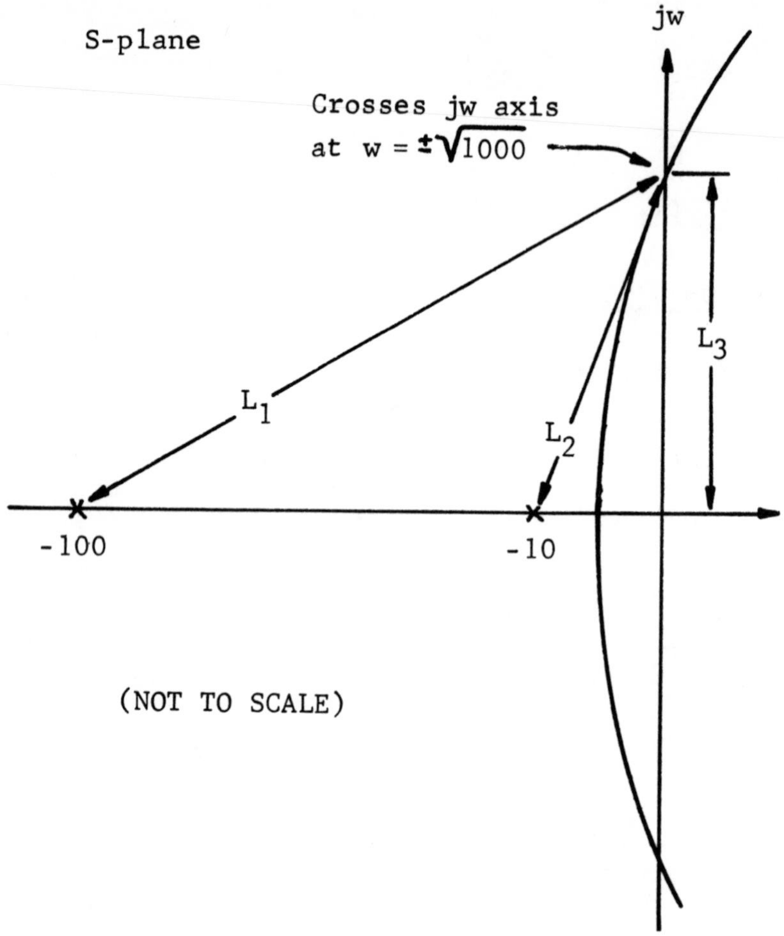

$$K_o = (L_1)(L_2)(L_3) = 11.0 \times 10^4$$

$$K = \frac{K_o}{1000} = 110$$

Thus for any value of K_o less than 110, the system would be stable.

CONTROL SYSTEMS 5

In a constant temperature bath, a bridge circuit is employed as the error detector, as shown in the figure.

> E is the temperature sensing element
>
> R is a fixed resistor of 100 Ω
>
> R_A and R_B are fixed resistors, the values of which are to be determined
>
> P is a potentiometer of 500 Ω for setting the bath temperature

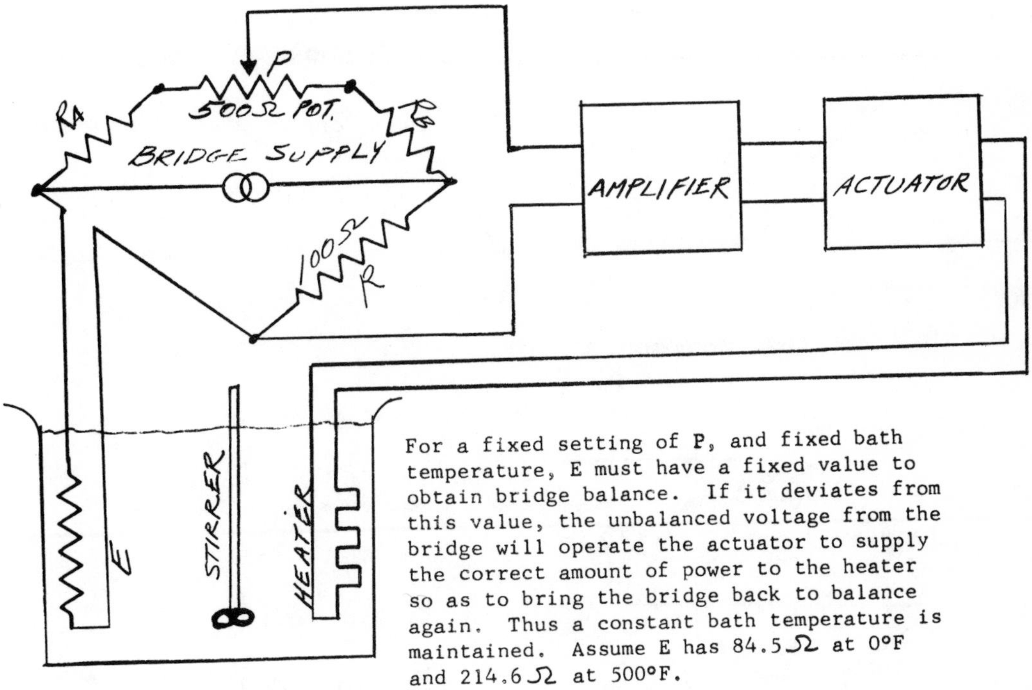

For a fixed setting of P, and fixed bath temperature, E must have a fixed value to obtain bridge balance. If it deviates from this value, the unbalanced voltage from the bridge will operate the actuator to supply the correct amount of power to the heater so as to bring the bridge back to balance again. Thus a constant bath temperature is maintained. Assume E has 84.5 Ω at 0°F and 214.6 Ω at 500°F.

REQUIRED:

Calculate the values of R_A and R_B so that by adjusting P, the bath temperature can be set at anywhere between 0° and 500°F.

SOLUTION

This involves solving a typical bridge balance problem with potentiometer "P" set to either of its extremes:

For 0° temperature

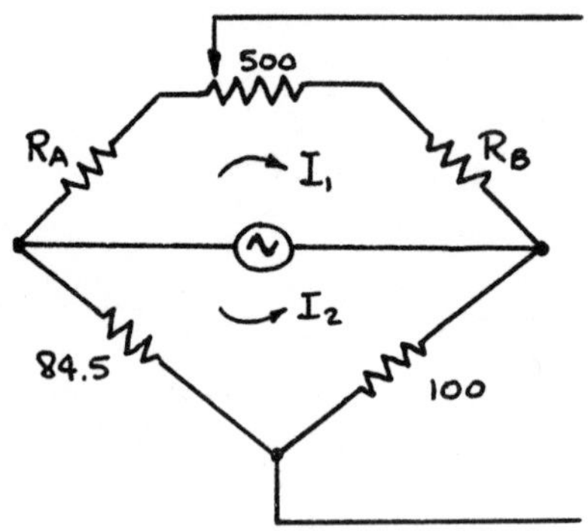

$$I_1 R_A = 84.5\, I_2$$

and

$$I_1(500+R_B) = 100\, I_2$$

dividing one equation by the other

$$\underline{R_A = 0.845(500+R_B)} \qquad (1)$$

And for 500° temperature

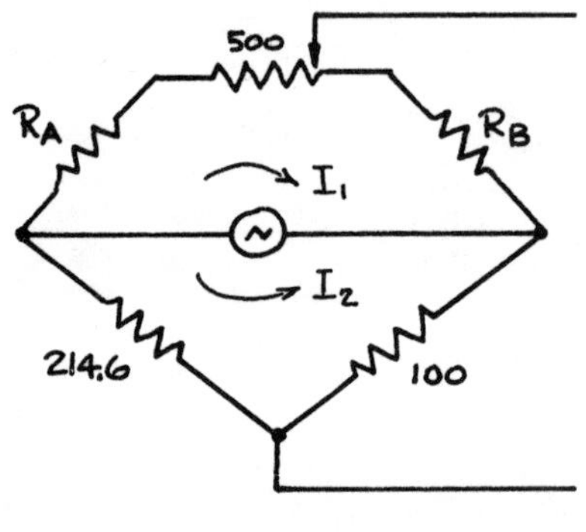

$$(500+R_A)I_1 = 214.6\, I_2$$

and

$$R_B I_1 = 100\, I_2$$

dividing one equation by the other gives

$$\underline{500+R_A = 2.146\, R_B} \qquad (2)$$

Then solving these two simultaneous equations (1 & 2) for R_A and R_B gives:

$$R_B = 710\ \Omega$$

$$R_A = 1022\ \Omega$$

CONTROL SYSTEMS 6

A viscous-damped servomechanism with proportional control stiffness K and damping coefficient B, is found to exceed the allowing tracking error by 10 times.

The latter is corrected by altering the stiffness, and by adding derivative control to provide the original degree of underdamping.

REQUIRED:

Find the new proportional control stiffness and the derivative action time required.

SOLUTION

A simple proportional control servomechanism might be represented as:

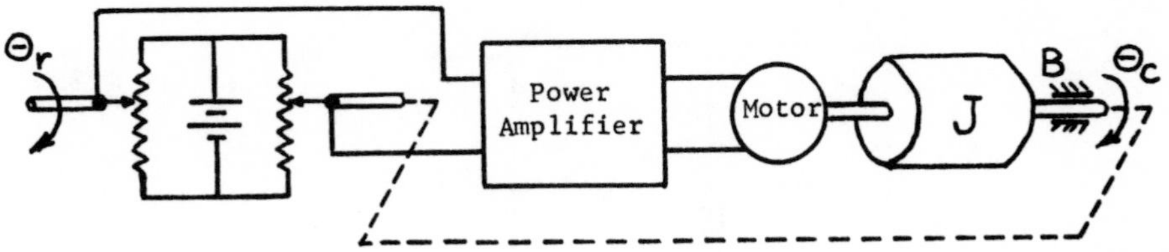

The equation describing this system could be written as:

$$K(\Theta_r - \Theta_c) = J\ddot{\Theta}_c + B\dot{\Theta}_c$$

(Equation assumes an oversimplified relationship where the motor torque is proportional to input voltage.)

And the Laplace transformed system then could be shown to be:

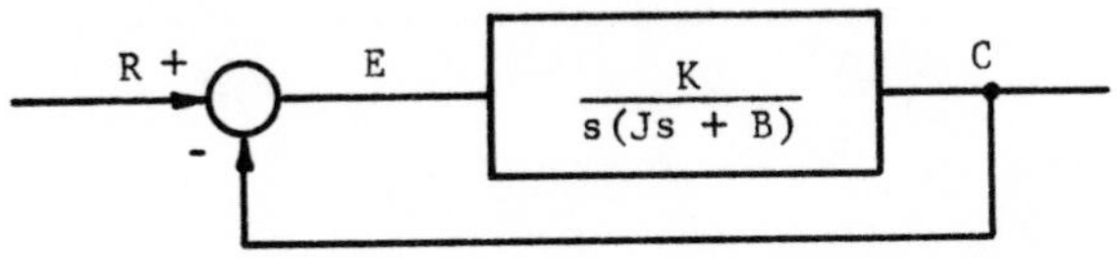

For ease of computing, let J = 1, then the system transfer function becomes:

$$G_{syst} = \frac{K}{s^2 + Bs + K} \equiv \frac{\omega_n^2}{s^2 + 2\zeta\omega_n s + \omega_n^2}$$

where $\omega_n = \sqrt{K}$

$$\zeta = \frac{B}{2\sqrt{K}}$$

The system tracking error (assuming a unit ramp input) then may be found using the final value theorem:

$$e\Big|_{t\to\infty} = \lim_{s\to 0} sE = \lim_{s\to 0} s\left[\frac{R}{1+G}\right] = \lim_{s\to 0} s\left[\frac{\frac{1}{s^2}}{1+\frac{K}{s(s+B)}}\right] \to \frac{B}{K}$$

And, the root locus will then be (assuming an underdamped system):

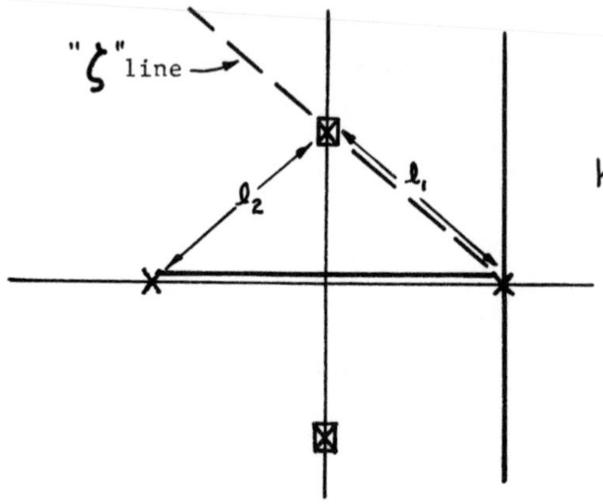

To reduce the system error to 0.1 of the original error, add an ideal derivative compensator:*

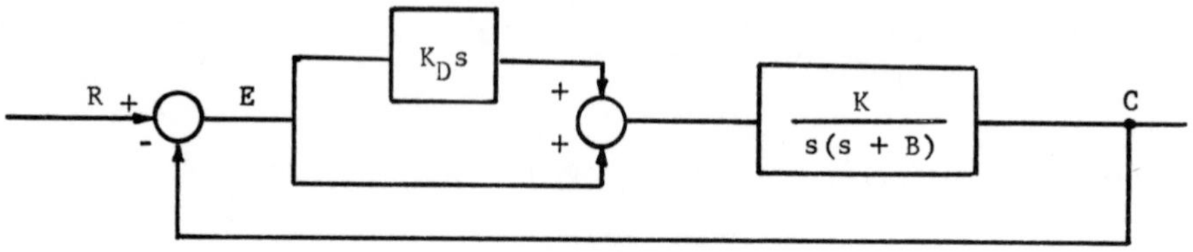

The new forward transfer function then becomes:

$$G_{fwd} = \frac{(1 + K_D s)K}{s(s + B)}$$

And the error again may be found using the final value theorem:

$$e\Big|_{t\to\infty} = \lim_{s\to 0} s\left[\frac{\frac{1}{s^2}}{1+\frac{(1+K_D s)K}{s(s+B)}}\right] \to \frac{B}{K}$$

*This portion of the problem requires a somewhat lengthly solution and a graphical presentation of the method would probably satisfy the exam grader.

To reduce the new error by a factor of 10 by increasing the stiffness, K, gives:

$$e_{new} = 0.1 e_{old} = \frac{B}{10K}$$

The new root locus will then have a zero located at $-\frac{1}{K_D}$ (yet to be determined).

Since the amount of damping is to remain the same (i.e., the "ζ" line to remain the same), the locus will be of the form:

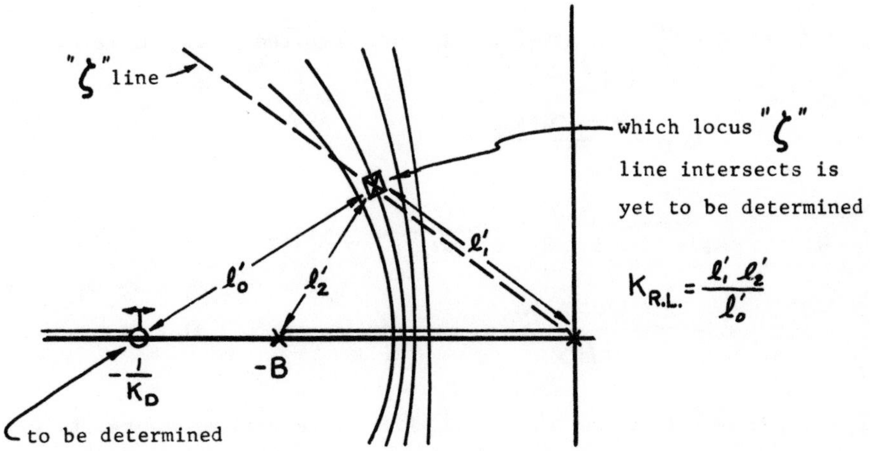

The requirement of error, being $K_{new} = 10 \times K_{original}$, and locating the new closed loop pole on the "ζ" line will give the relationship of:

$$K_{root\ locus} = \frac{\ell_1' \ell_2'}{\ell_0'} = K_D(10K) = K_D 10 \ell_1 \ell_2$$

Using this relationship, one may, by trial and error techniques, find the value of K_D.

However, a more straight forward technique not using root locus methods, will allow us to approximate K_D directly (this will be approximate as the zero in the closed loop response will alter the damping).

Consider:

$$G_{syst} = \frac{(10K)(1 + K_D s)}{s^2 + Bs + K_D(10K)s + (10K)} \equiv \frac{\omega_n^2(1+K_D s)}{s^2 + 2\zeta\omega_n s + \omega_n^2}$$

where $\omega_n = \sqrt{10K}$

$$2\zeta\omega_n = B + 10K\,K_D$$

$$\zeta = \frac{B + 10K\,K_D}{2\sqrt{10K}}$$

But the new "ζ" and original "ζ" are required to be the same.

Giving:

$$B = \frac{B + 10K\,K_D}{\sqrt{10}}$$

As an example, let K = B = 2 and "ζ" = 0.707

Then

$$2 = \frac{2 + (10)(2)K_D}{\sqrt{10}} \qquad \text{Giving } K_D = 0.216$$

Or, the zero (for the root locus plot) is located approximately at $\frac{-1}{K_D}$, which gives -4.6 for the numbers previously used.

Therefore the new stiffness should be 10 times the original and the "time constant" of the derivative compensator should equal K_D.

CONTROL SYSTEMS 7

A servo system for the positional control of a rotatable mass is stabilized by means of viscous damping. The amount of damping is less than that required for critical damping.

REQUIRED: Calculate the amount (percent) of the first overshoot, if the input member is suddenly moved to a new position, the undamped natural frequency is 5 Hz, and the viscous friction coefficient is a fifth of that required for critical damping.

SOLUTION

Assume an ideal controller with torque directly proportional to error, E, (T = KE), and that system is of 2nd order:

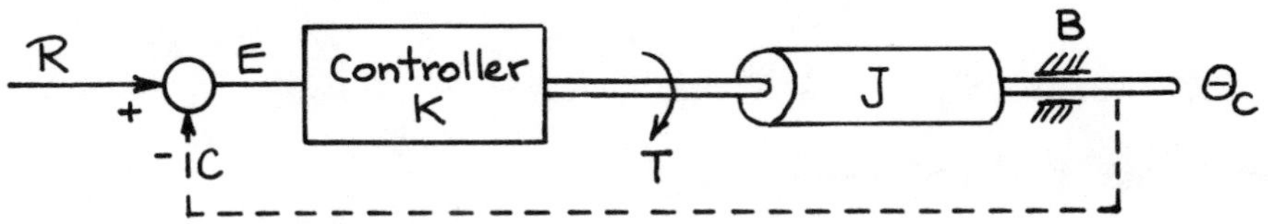

$$T = KE = K(R - C)$$

$$T = (JS^2 + BS)\Theta$$

$$G_{SYST} = \frac{K}{JS^2 + BS + K} = \frac{K/J}{S^2 + \frac{B}{J}S + K/J}$$

$$\equiv \frac{\omega_n^2}{S^2 + 2\zeta\omega_n S + \omega_n^2}$$

ω_n = natural frequency = $2\pi f_n = 2\pi 5$ = constant

$2\zeta\omega_n = B/J$, $\therefore \zeta \propto B$ since both K and J are constants.

For critical damped system, $\zeta = 1$.

For $B = 1/5\, B_{critical}$, $\zeta = 0.2$

From "standardized" 2nd Order Curves:

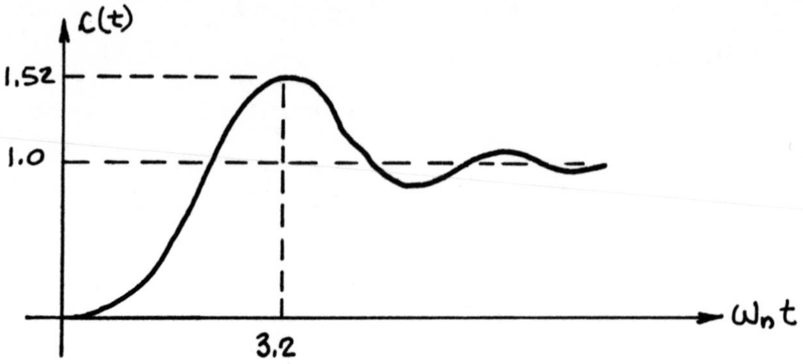

∴ % overshoot = 52%

and time-to-first-peak, t_p

$$t_p \approx \frac{3.2}{\omega_n} \approx 0.1 \text{ seconds}$$

8

Digital Computers

The subject material under the heading of digital computers which could be on the examination is extremely broad. In Chapter I the technical specialities as set forth by NCEE (1986) are fully listed; that portion of the list concerning digital computers will be repeated here:

> 4. COMPUTERS: Four (4) problems relating to digital hardware analysis and design (combinational and sequential); integrated circuit digital devices (counters, multiplexers, ROM, PLA, etc); microprocessors and applications; programming (in higher level language, such as BASIC, FORTRAN, or PASCAL) for dynamic analysis and design; digital computer simulation (such as CSMP); network planning; data communications.

Obviously, this amount of material would take several volumes to review. Thus only the direction of study along with very limited review material (for those who have not been active in this area for some time) for the purpose of "jogging" one's memory will be presented in this chapter.

DIGITAL LOGIC Assuming the function of logic gates are understood, one needs to start with the review of logic design minimization by use of the Karnaugh map. From a problem statement, a truth table is formed from the natural binary code. As an example, assume that "from three ON/OFF switches, A, B, and C, it is required to have an output signal that is ON (high) if switches A and C are ON, or if B and C is ON but only if the digital number that represents the signals from A, B, and C is odd; also, if these three signals represents the digital number four, we don't care whether the output is ON or OFF." The truth table is given and the minimal sum of products expression is found from the grouping in the Karnaugh map

	m	A	B	C	f
	0	0	0	0	0
(odd)	1	0	0	1	0
	2	0	1	0	0
(odd)	3	0	1	1	1
	4	1	0	0	X
(odd)	5	1	0	1	1
	6	1	1	0	0
(odd)	7	1	1	1	1

$f(A,B,C) = \sum m(3,5,7) + dc(4)$

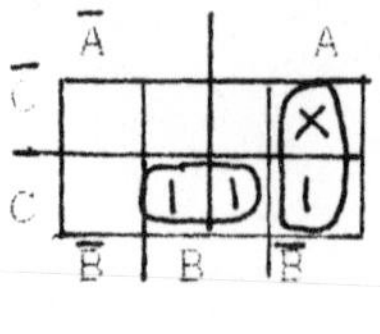

$f = BC + A\bar{B}$

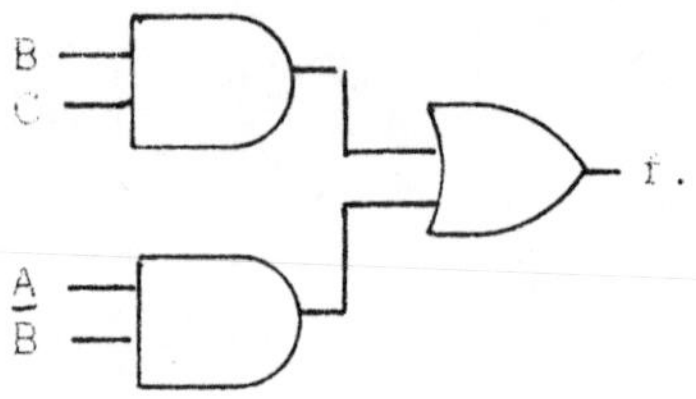

(Another minimal solution is possible, f = BC + AC.)

For decoding a group of signals, recall that a standard DECODER may also be used instead of the above network to give the desired output signal as follows. (As a help in starting your review,

you should first check the internal circuit of a typical 1 out of 8 decoder; or, even better, derive the circuit from a truth table.)

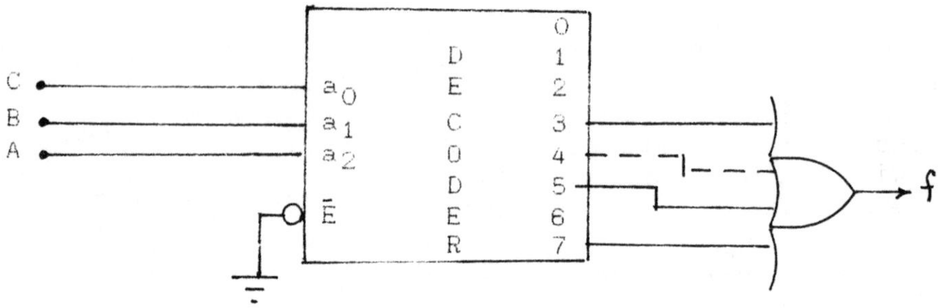

Recall the function of a multiplexer (MUX); it selects a particular line from several lines and passes the information directly to its output. Consider the following circuit equivalents:

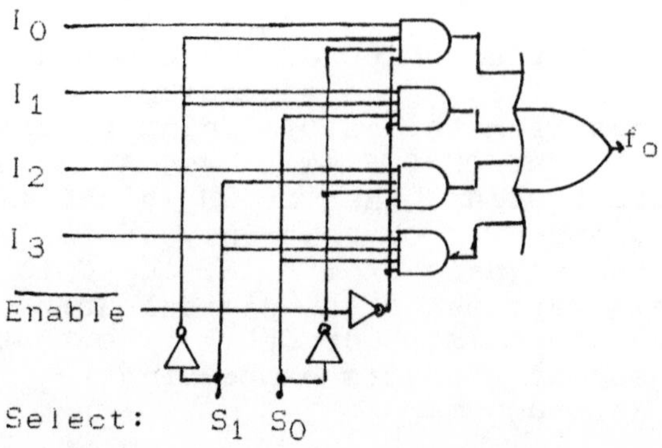

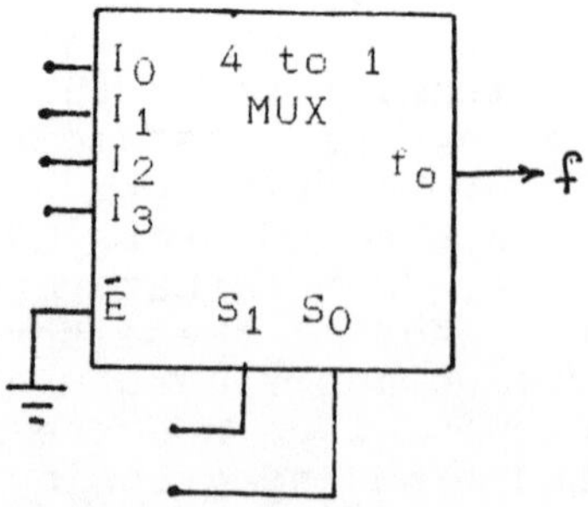

Consider the use of the MUX as decoder by reference to a truth table and a Karnaugh map. Assign as many variables as possible to the select lines (refer to previous example); one then could assign the other variables through logic combinations to the MUX inputs. Consider the previous example

m	A=S	B=S	C	f
0	0	0	0	0
1	0	0	1	0
2	0	1	0	0
3	0	1	1	1
4	1	0	0	x=0
5	1	0	1	1
6	1	1	0	0
7	1	1	1	1

	$\bar{S}_1$	S_1
$\bar{S}_0$	I_0	I_2
S_0	I_1	I_2

	$\bar{S}_1$	S_1
$\bar{S}_0$	0	C
S_0	C	C

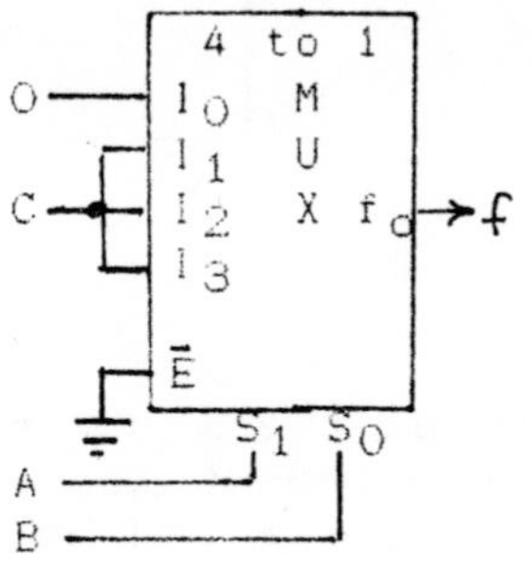

For a Read Only Memory (ROM), it should be recalled that binary values may be stored in a matrix form and could be represented by either a voltage (1) or no voltage (0) and this voltage may be obtained from the voltage drop across resistor. Consider the following oversimplified circuit where the current is directed by diodes; missing diodes mean no current flows. (Note that the crossing wires are nontouching.) By selection the address, or rotary switch number, the particular horizontal wire is addressed; then current may or may not flow in the vertical data output wires.

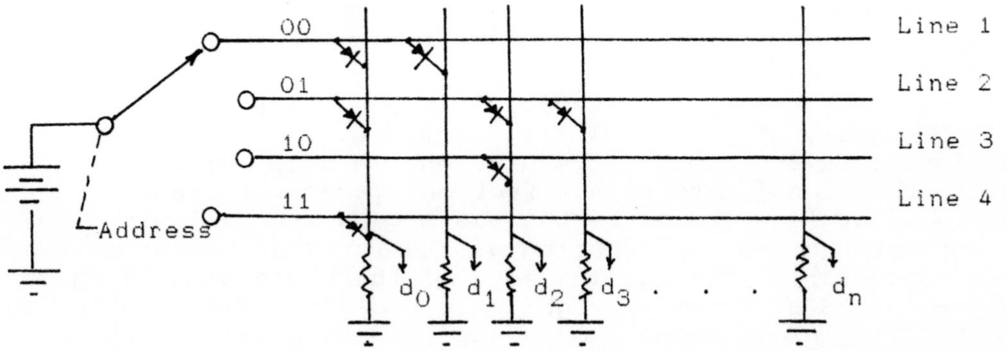

Of course the mechanical switch is replaced with a decoder (which is normally included in the ROM package); The address is in the form of the binary code such that if one new the number of address lines (al), the number of horizontal lines is given by 2al. Also note the number of data output lines is independent of the number of address lines. Again, refer to the previous example; we may use addresses as the signal lines and connections by the diodes will yield the desired output.

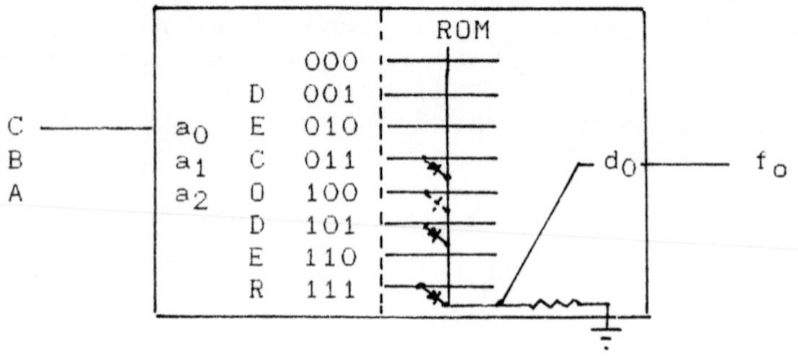

Normally many other output data lines are available for stored data; this is an oversimplified example.

One of the most important elements of a digital system is the flip-flop; this element usually may be used as a latch or may be edge triggered by an external signal. The basic circuit may be made up from a pair of NAND or NOR gates. Consider the NOR pair and its flip-flop representation. Call one input SET (S), the other RESET (R), one output Q, and the other Q; this f-f is called a latch.

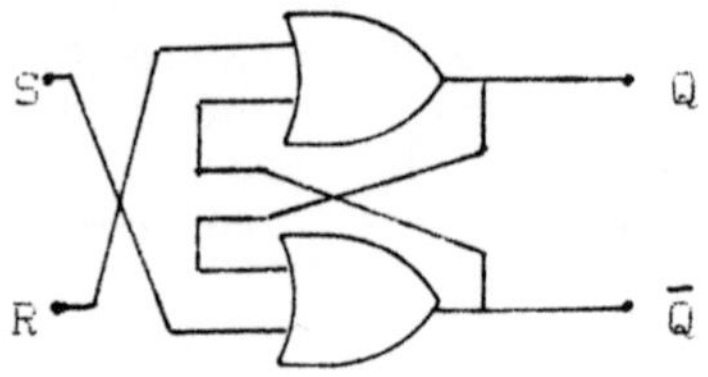

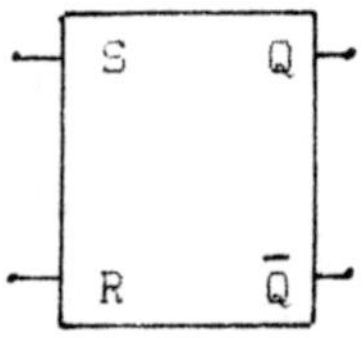

If, on the other hand, the signals S and R are withheld by a pair of AND gates, the f-f will stay is whatever state it happens to be in unless the signal C is also on. A step further would be to allow the signal C to have effect only when the leading edge of a pulse is applied (think of this as a capacitor allowing the wave front through and the function of the resistor is the discharge the capacitor). The leading edge of the C signal will cause any signals on S or R get through; this C signal is thought of as the controlling clock signal (only the leading edge).

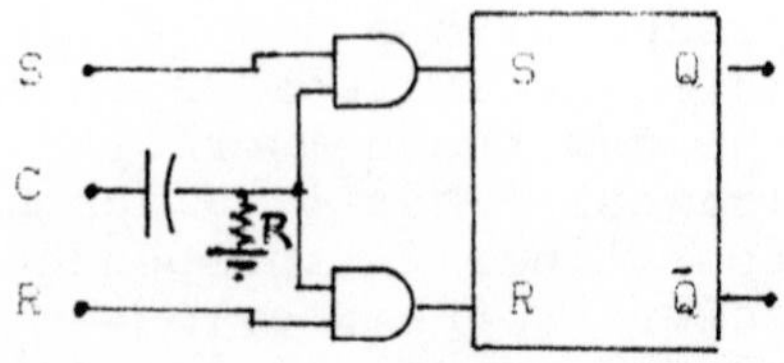

Latch f-f

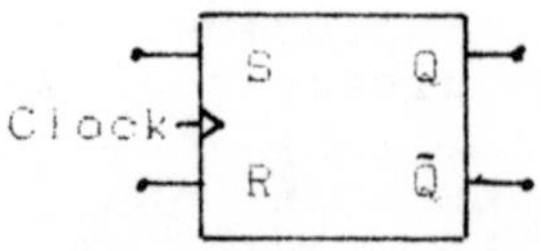

Edge Triggered (leading) f-f.

The most popular versions of these edge triggered f-f's are the "T", "D", "J-K", and the master-slave units. Recall that the time delay through the f-f allows us to use these devices in synchronous circuits (i.e., all driven by a common clock). Also recall that the truth table for the S-R and J-K f-f's are the same EXCEPT the inputs on the S-R f-f should never have 1's at the same time while the J-K may both have 1's on their inputs.

	S	R	Q	Q_{next}
	0	0	0	0
	0	0	1	1
	0	1	0	0
	0	1	1	0
	1	0	0	1
	1	0	1	1
Not Allowed	1	1	0	?
	1	1	1	?

J	K	Q	Q_{next}
0	0	0	0
0	0	1	1
0	1	0	0
0	1	1	0
1	0	0	1
1	0	1	1
1	1	0	1
1	1	1	0

As an example consider the following configuration; assume all Q's are cleared to start (i.e., set to zero).

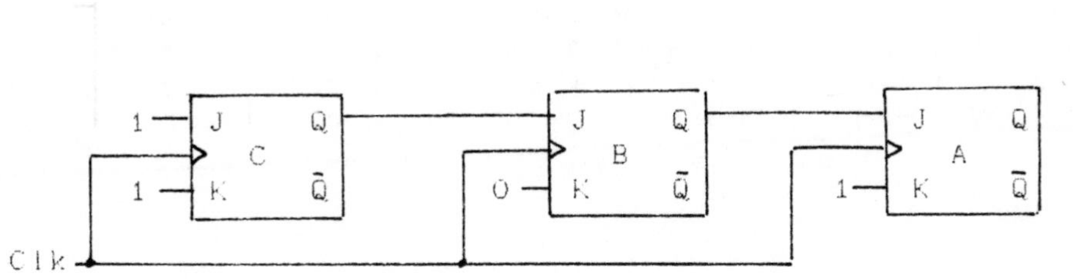

The timing diagram is shown below. It should be noted that if one neglected the time delay through the f-f's, the timing diagram would not take into account the correct signals "ready and waiting" on the respective inputs when the steep wave front of the clock signal arrives.

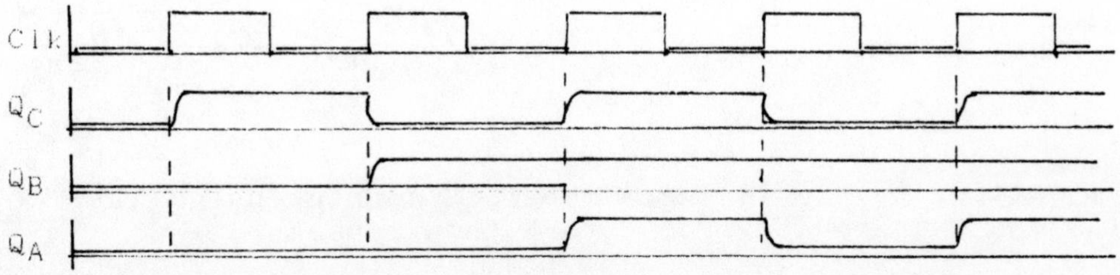

SEQUENTIAL COUNTER

Consider the design of a sequential "odd ball" counter that has a sequence of, say, 0,1,3,2,0, . . . repeat. Since four different values are represented, then only two flip flops are needed (i.e., 2n=4, then n=2). Assume only one J-K type and one R-S type f-f is available; a state table may be developed along with a state diagram to help one formulate the necessary logic. Designate one f-f as the most significant bit (MSB), say the J-K type f-f, and the other f-f as the least significant bit (LSB). The state table may be completed from the truth table of these two types of f-f.

m	PS		NS		J-K		S-R	
	MSB	LSB	MSB	LSB	J	K	S	R
0	0	0	0	1	0	X	1	0
1	0	1	1	1	1	X	X	0
3	1	1	1	0	X	0	0	1
2	1	0	0	0	X	1	0	X
0	REPEAT - - - - - - - -							

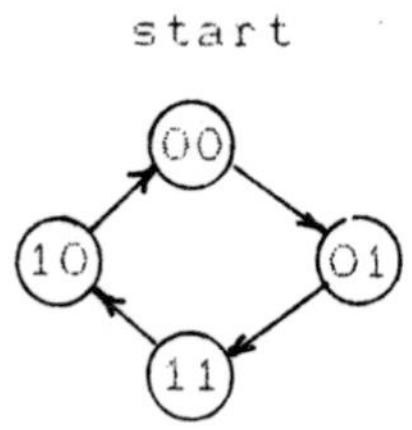

where PS is the present state and NS is the next desired state and the J-K f-f is designated as A and the S-R ff as B. Karnaugh maps may be used for each input of each f-f to form the circuit logic; then a timing table is sketched to check for the correct operation:

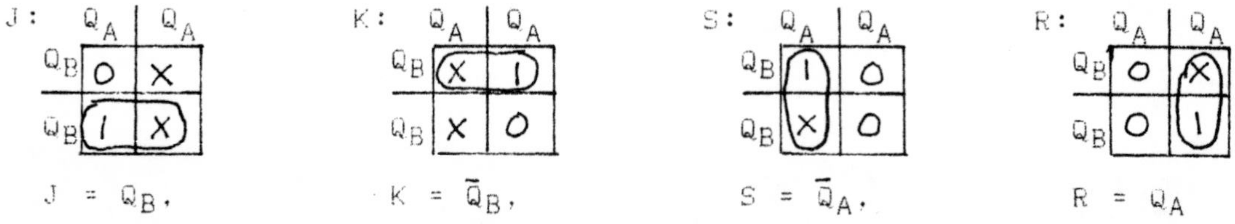

$J = Q_B$, $K = \bar{Q}_B$, $S = \bar{Q}_A$, $R = Q_A$

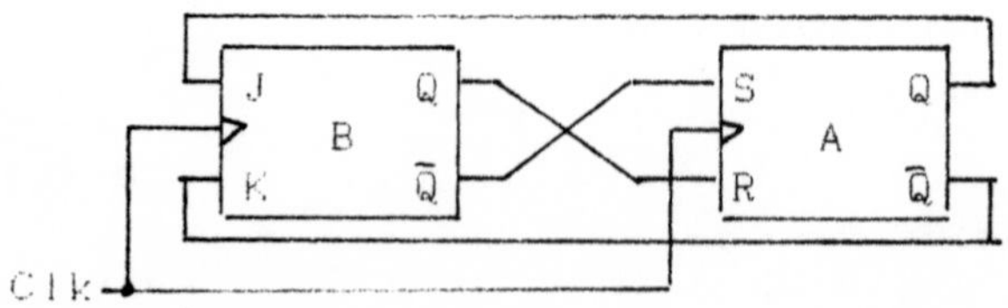

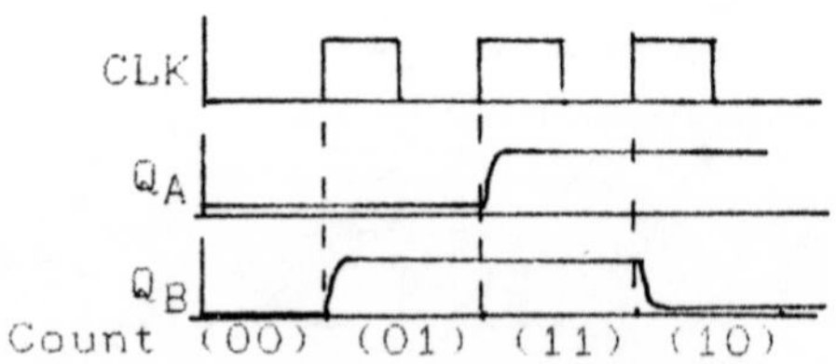

For practice, try designing a modulo 5 counter using all J-K f-fs (i.e., one that counts from 0 to 5 then repeats).

Final solution:

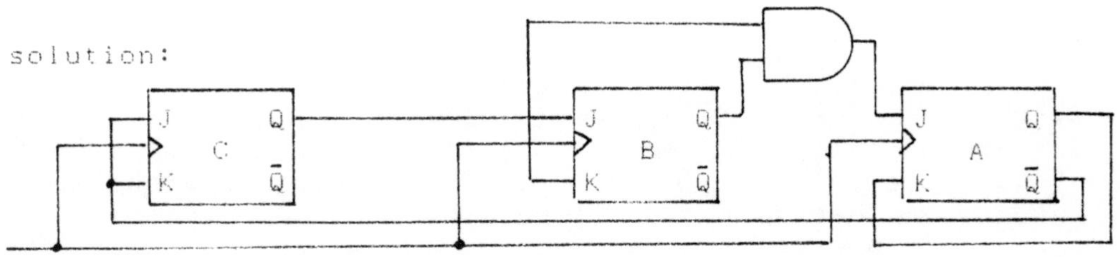

MICROPROCESSORS This subject area is much too lengthy to be included in a short review such as this. Usually one is either versatile in this area or has never been involved in the actual design of a problem solution using this device. To attempt to study this subject for the first time for the sole purpose of being able to answer questions is probably not the best use of one's time. For those who still would like a fast "run through" of this subject, but have no background in it, an excellent text written at the community college level is Malvino's DIGITAL COMPUTER ELECTRONICS, McGraw-Hill, 1977. His chapters on SAP-1, SAP-2, and SAP-3 (Simple As Possible microcomputer) is perhaps the best and simplest introduction of the subject by any author.

HIGHER LEVEL LANGUAGE AND SIMULATION Of the possible languages suggested, probably BASIC is the easiest to review in a short amount of time. This is most easily done by refreshing one's own memory with the various calls; here the suggestion is to obtain a "quick reference" booklet and review the purpose and syntax of the BASIC commands, statements, functions, and variables. Of course if one is more familiar with FORTRAN or PASCAL then a choice of one of these would be more appropriate to review; for those having an interest in simulation, FORTRAN is usually the base language and the preferred one to review.

The subject of SIMULATION is also a very broad subject with many different programs available; a very broad categorization of the subject would be to separate simulation into continuous or discrete event types (discrete events type is not discussed in these few pages). The so-called "old standby" of continuous simulation is CSMP and is still one of the most popular and contains most of the elements of the newer programs such as CSSL, MATRIXx, CC, TUTSIM, and ACSL. Developed especially for the IBM-PC is one called PCESP; for more information on this program, please refer to the text by Shah, ENGINEERING SIMULATION TOOLS AND APPLICATIONS USING THE IBM PC FAMILY, Prentice Hall, 1988. Although a short review of THE CSMP program will be attempted, an excellent and also leading text on this subject is Speckhart & Green's A GUIDE TO USING CSMP - THE CONTINUOUS SYSTEM MODELING PROGRAM, Prentice-Hall, 1976, is available for a more thorough study.

CSMP is more like using an analog computer rather than writing programs in one of the higher level languages; one may use CSMP without reference to FORTRAN but, for more advance usage, a background in FORTRAN is most helpful. As with all of these simulation packages, prepackaged algorithms for integration and many other operations are available. As an example, several different integration approximation algorithms are at one's disposal depending on which "fits" a particular problem. If no integrating method is specified, the program automatically default to a variable step type (such as the Runge-Kutta or Milne method). The default routine is usually more than adequate for almost all but the more specialized type of problem (i.e., widely separated time constants, discrete time delay sample and hold functions, etc.); for sample and hold simulations, a fixed step algorithm should be used.

CSMP program statements usually fit into three categories; these categories are the data statements, structure statements and control statements. Generally the structure of CSMP has three segments to it; these are the initial, the dynamic, and the terminal segments. Initial conditions and direct calculations, such as the area of a circle or whatever, may be placed in the initial segment, while the dynamic segment includes anything involving iterative computing of the system such as the describing differential equations. The terminal segment may include control statements, plotting algorithms or routines, output statements, and any timing statement. Arithmetic operations use the same symbols as in FORTRAN. Automatic sort routines are incorporated and may be assumed unless specifically called out with a nosort statement; nosort would be used for special procedural subroutines such as branching and counting. To discuss all but the simplest concepts, one needs to know the version of CSMP that is available and also have the necessary manual defining the various block and integration routines.

A very simple example will be given to indicate the method. However, first a few definitions and blocks will be given:

Program Statement	Function
Y = INTGRL(IC,X)	y = xdt + IC, X → [1/s] → Y
Y = REALPL(IC,TC,X)	TCy + y = x, X → [1/(TCs+1)] → Y
Y = STEP(T)	y = 0; for t<T, y = 1; for t>T 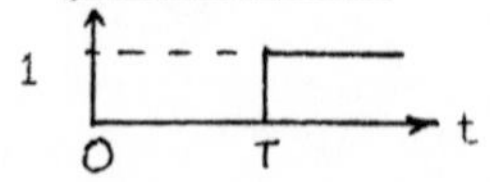
Y = LIMIT(N_1, N_2, X)	y=N_1; for x<N_1; y=N_2; for y>N_2 y=x; for N_1<x>N_2 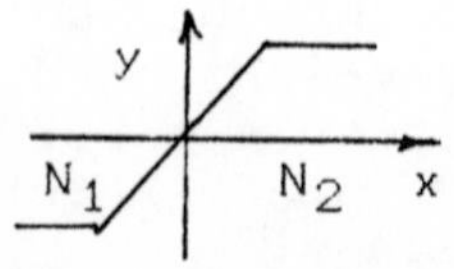

Assume one wishes to simulate the following nonlinear block diagram:

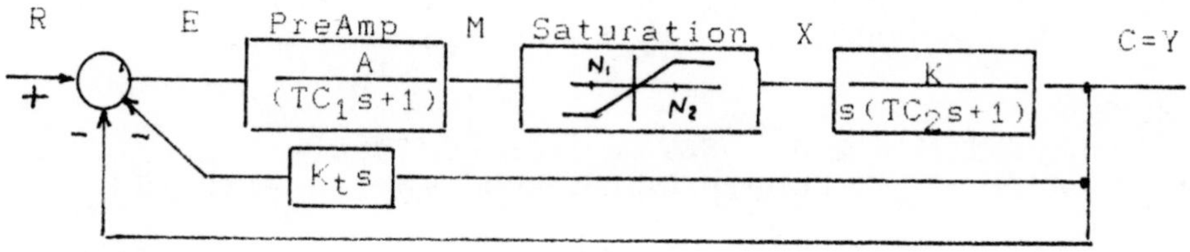

It is recommended to avoid differentiation both for analog and digital simulation; therefore, it is convenient to rearrange the above problem as follows

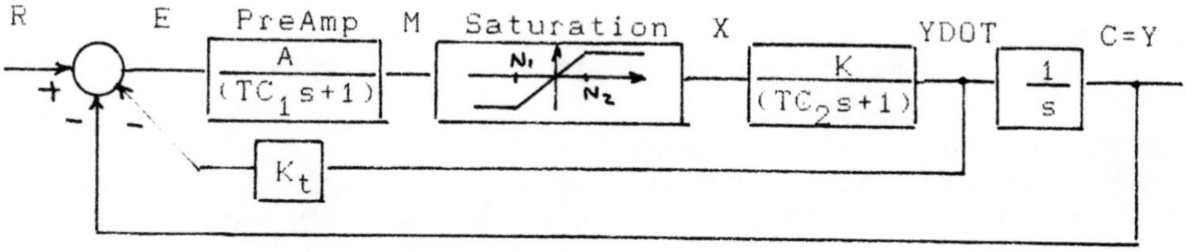

The CSMP simulation program is

```
TITLE  PROGRAM TO SIMULATE FEEDBACK SYSTEM
   INITIAL
      CONSTANT A=5.0, K=2.0, TC1=0.005, TC2=0.5, N1=-1.0, ...
               N2=1.0, KT=0.025
   DYNAMIC
         E = STEP(0) - K*YDOT - Y
         M = REALPL(0,TC1,E)
         X = LIMIT(N1,N2,M)
         YDOT = REALPL(0,TC2,X)
         Y = INTGRL(0,YDOT)

*COMMENT  THIS COMPLETES THE DYNAMIC PORTION

    TERMINAL
         TIMER  FINTIM = 2.5, PRDEL = 0.1, DELT = 0.01
         PRINT Y, E, M
         END
         STOP

*COMMENT   PROGRAM FINISHES IN 2.5 SEC., THE PRINTOUT IS EVERY
*COMMENT   PRDEL SECONDS, BUT THE MINIMUM STEP SIZE OF THE
*COMMENT   INTEGRATION INTERVAL IS TO BE DELT
*COMMENT   IF DEL IS NOT SPECIFIED, THE DEFAULT IS
*COMMENT   DEL=FINTIM/100
```

Since the integration method is not specified, the default is a variable step size Runge-Kutta method. If the system contained a sample and hold portion for z-transforms, then one would insert a statement to the effect that a fixed step integration method would be used; also the DEL would have to be some sub-multiple of the output interval (the first step is 1/16th of PRDL or OUTDEL). As an example

```
     METHOD RKSFX
     TIMER FINTIM = 2.0, DELT = 0.01, OUTDEL = 0.16
*COMMENT  RKSFX IS A FOURTH ORDER RUNGE-KUTTA WITH FIXED INTERVAL
```

To review for any of the simulation programs, one, of course, needs the manual and then go over two or three of the integration routines besides the various block calls.

DIGITAL COMPUTERS 1

This is a 10-question multiple choice problem. It may be longer and more detailed than might be expected on the actual exam.

Situation: An Off-ON type sensor (OFF=0 volts, ON= 5 volts) is used to measure the vibration level in a rocket motor (ON if vibration level exceeds a preset threshold); when ON, a warning device is activated. However, because overly high vibrations can sometimes change the preset value of the sensor, it is proposed to replace the one sensor with three (the line of reasoning being that if only one sensor is on, it is probably out of correct calibration). A logic circuit is designed such that the warning device should be activated only if two or more sensors are ON. Assume the sensors are A, B, and C.

QUESTION 1 The nonminimal function (i.e., canonical form) for the warning device to be activated is:

a) $f = \bar{A}B\bar{C} + A\bar{B}C + \bar{A}BC + AB\bar{C} + \bar{A}\bar{B}\bar{C}$
b) $f = \bar{A}BC + A\bar{B}C + AB\bar{C} + ABC$
c) $f = \bar{A}\bar{B}C + A\bar{B}C + A\bar{B}\bar{C}$
d) Correct function not given since only two sensors need to be activated.
e) All functions listed are correct since only two sensors need to be activated.

SOLUTION Use the truth table to form the function on two or more 1's for the sensors.

m	A	B	C	f
0	0	0	0	0
1	0	0	1	0
2	0	1	0	0
3	0	1	1	1
4	1	0	0	0
5	1	0	1	1
6	1	1	0	1
7	1	1	1	1

f(to activate alarm) $= \bar{A}BC + A\bar{B}C + AB\bar{C} + ABC$

Answer is (b).

QUESTION 2

The minimal sum of products expression for the warning device to be activated is:

a) $f = AB + AC + BC$
b) $f = \bar{A}BC + A\bar{C}$
c) $f = AC + A\bar{B} + B\bar{C}$
d) $f = A\bar{C} + \bar{B}C$
e) Correct function not given since only two sensors need to be activated.

SOLUTION

Form a Karnaugh map for $f = \sum m(3,5,6,7)$.

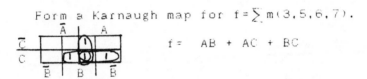

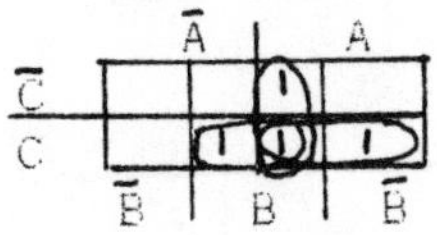

$f = AB + AC + BC$

Answer is (a).

QUESTION 3

Another logic circuit is designed such that a second warning device will come ON if all three vibration sensors do not agree. The minimal product of sums expression for this second warning device is given by:

a) $f = ABC + \bar{A}B\bar{C} + A\bar{B}C$
b) $f = (A+\bar{B})(\bar{B}+C)(A+C)$
c) $f = (A+\bar{C})(\bar{B}+C)(\bar{A}+B)$
d) $f = (A+\bar{B}+C)(\bar{A}+B+\bar{C})$
e) Correct function not given because second warning light will always be ON.

SOLUTION From the same type of truth table as Q-1, form the function when A, B, and C agree (i.e., terms 0 and 7).

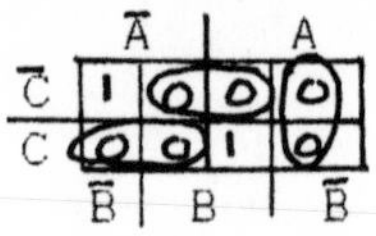

$\overline{f}$(activate 2nd alarm) = $\bar{A}C + B\bar{C} + A\bar{B}$

$\overline{\overline{f}}$(" " ") = $\overline{\bar{A}C + B\bar{C} + A\bar{B}}$

f(" " ") = (A+C)(B+C)(A+B)

Answer is (c).

QUESTION 4

It has been determined that vibration spikes that exceed the preset values of the sensors that only last for 50 ms or less may be neglected and should not turn on the first warning device. A logic circuit is designed to be inserted in the line coming from the previously designed logic (i.e., the output activating circuit) and the warning device. This new circuit should not allow the activating signal to pass to the warning device for the first 50 ms of a high vibration and then reset itself. One such design is as follows (it is known that a J-K latch "flips" if the "J" terminal exceeds 2.9 volts and its input impedance is negligibly high before triggering):

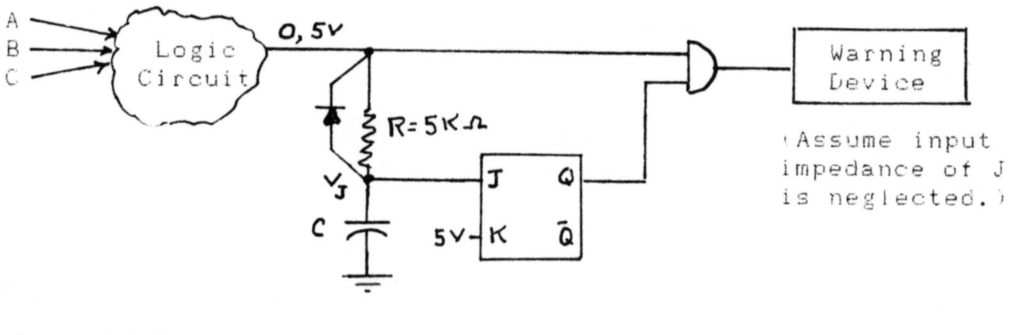

For the RC circuit, a value of 5K ohms is selected for R; what should be the nearest value of C?

a) C = 5 ufd
b) C = 10 ufd
c) C = 50 ufd
d) C = 100 ufd
e) Not possible for circuit to operate (for any value of C) within specifications since "K" is held at 5 volts.

SOLUTION $V_J = 2.9 = 5(1-e^{-(1/RC)t}) = 5(1-e^{-x})$, $-x = 0.867$
$0.867 = (1/RC)t$, for R=5K, C=11.5 ufd

Answer is (b).

QUESTION 5

The circuit for part 4) proves to be too inaccurate and it is decided to use the following counting circuit instead, what is the nearest value of clock frequency that should be used?

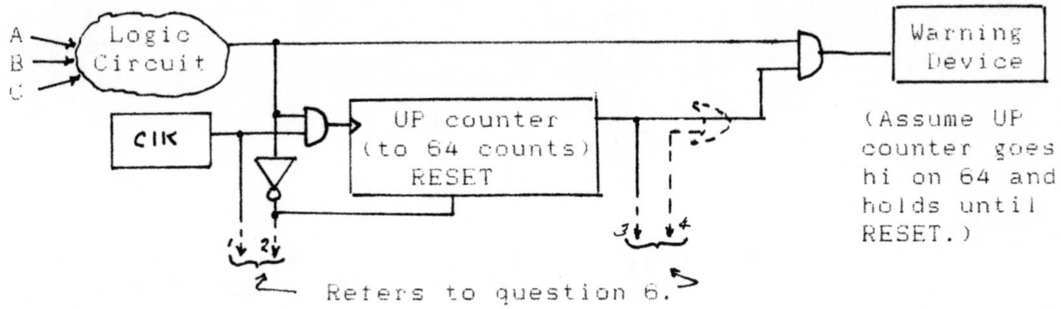

a) 12 Hz
b) 1.2 Khz
c) 64 KHZ
d) 50 MHz
e) 3.2 KHz

SOLUTION 64 counts must occur in 50 ms or less, or:
64/50ms=1.28×10^3, clock=1.28 KHz

Answer is (b).

QUESTION 6

Although the circuit of question 5 satisfies the new specifications, it is again later decided that if a high vibration has a duration of less than 50 ms (such that the warning device is not activated), then a window of another 50 ms (for low vibrations less than the preset value) should pass before another delay of 50 ms may be tolerated (i.e., warning device should be activated should another high vibration occur before the 50 ms has elapsed). The following circuit with another 64 count counter is added (at the dashed lines) to that of question 5:

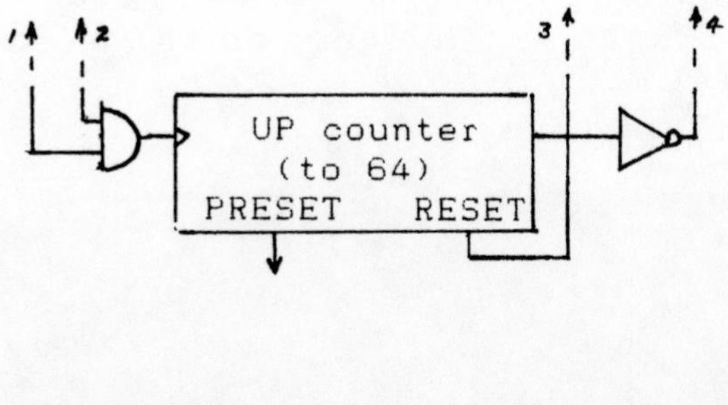

a) For correct operation PRESET should be tied hi.
b) " " " " " " " lo.
c) " " " " " " left floating.
d) " " " " " " to output of right hand AND gate.
e) Circuit is incorrect and can't satisfy these new specifications.

SOLUTION Sketching a timing diagram will show that the added counter output should always start hi otherwise warning device will stay hi.

Answer is (a).

QUESTION 7

Refer to original problem with only the one sensor but whose output is an analog voltage (rather than OFF/ON) such that the output voltage is linear from 0 to 5 volts (with 0 corresponding to no vibration and 5 volts being the highest vibration expected. A standard 8-bit successive approximation type A/D converter is used to interface a small digital computer. For a vibration level corresponding to mid-range (i.e., 2.5 volts), the nearest value of the digital out (after the end-of-conversion signal is sent) is nearest to which of digital outputs expressed in hexadecimal values:

a) 55
b) F7
c) 7F
d) 127
e) 2.5

SOLUTION Mid range corresponds to $(127)_{10}$. Note that $(255)_{10}$ is for full scale for an 8-bits converter which, in turn, corresponds to $(7F)_{16}$.

Answer is (c)

QUESTION 8

Again it is decided to activate an alarm device if the output of the A/D converter exceeds a given value. The design of the circuit that is connected to the output lines of the A/D converter is such that the MSB and the next MSB lines are taken to an AND gate and the output this gates turns on the alarm. The analog voltage output of the sensor is approximately which of the following voltages to just turn on the alarm?

a) 0.5 volts
b) 0.75 "
c) 1.8 "
d) 3.8 "
e) 4.8 "

SOLUTION The MSB corresponds to 128 and the next MSB is 64, and 128+64=192, therefore (192/255)5.0=3.75 volts.

Answer is (d).

QUESTION 9

Assume that in addition to the alarm circuit of question 8, an extra EMERGENCY ALARM is to be activated if the analog voltage level from the sensor exceeds 3.92 volts. The circuit chosen is that of question 8 along with another AND gate (whose output will activate this new alarm); one of its inputs will come from the output of the AND gate (of question 8), and other will come from:

a) the output of the sensor.
b) the output of the LSB line from the converter.
c) " " " " MSB " " " " .
d) " " " " next LSB line from the converter.
e) " " " " 3rd next LSB line from the converter.

SOLUTION For the configuration of question 8 the "weight" of the two lines are equivalent to $(192)_{10}$ or 3.75 volts, therefore 3.90 -3.75=0.15 volts, or (0.15/5.0)255=7.65 (nearest integer value=8). Therefore the line that represents 2^3 is correct.

Answer is (e).

QUESTION 10

If the desired A/D converter (for question 8) is an 8-bit type, but only a 16-bit type is available, then one may substitute the 16-bit converter by:

a) Doubling the clock rate.
b) Using the MSB eight bits of the data bus and grounding the lower eight bits (assuming the same reference voltage).
c) Using the lower eight bits and grounding the upper eight bits (assuming the same reference voltage).
d) Using the lower eight bits and grounding the upper eight bits (but doubling the reference voltage).
e) Using the middle eight bits and grounding the others (assuming the same reference voltage).

Answer is (b).

DIGITAL COMPUTERS 2

Situation: A tentative design for a control system to position a radar antenna has been designed. The system has been designed with out any feedback compensation (such that minimal cost and weight factors may be achieved); however, if need be, this feedback compensation could be added in by a value of K_T. Before the system is built, it is decided to test the design by computer simulation to check the performance of the system without any feedback compensation (i.e., $K_T=0$). The tentative design block diagram is given as:

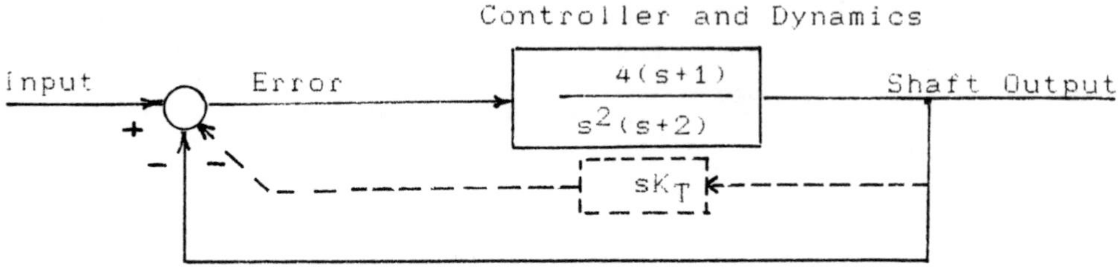

SOLUTION

Of the many high level simulation languages available, it is decided to use CSMP with the default method of integration. Also, since the programmer is not familiar with optimization techniques of analysis for the feedback compensation, the solution for the output vs. time for a step input will be programmed for K_T equal to zero; then if the results are not satisfactory, K_T will be set in and incremented until the desired results are achieved.

Because of the possible use of the derivative function used for the feedback compensation, the block diagram is first rearrange to avoid this operation.

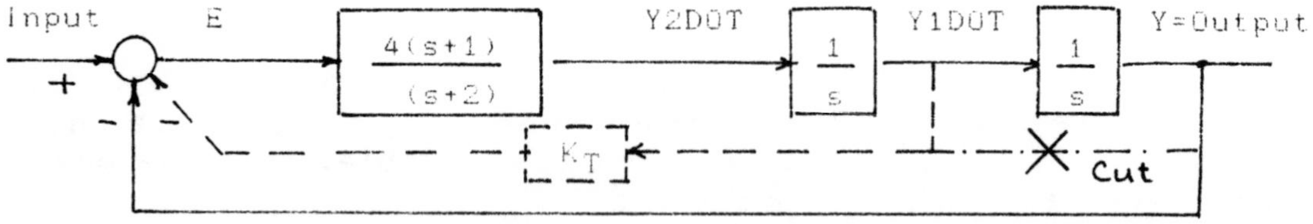

Or, for programming conceptualization, again the diagram is modified as:

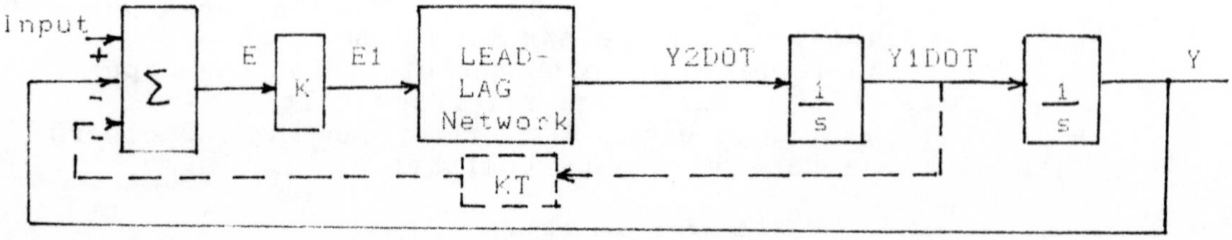

It should be noted from a CSMP manual that a lead-lag network is given by:

$$\frac{Y(s)}{X(s)} = \frac{P1s+1}{P2s+1} \quad \text{or} \quad P2(dy/dt)+y = P1(dx/dt)+x.$$

Then $\frac{4(s+1)}{s+2} = \frac{2(s+1)}{0.5s+1}$

Thus, one possible simulation program could be:

```
LABEL  SIMULATION OF AN ANTENNA POSITION CONTROL SYSTEM
INITIAL
   CONSTANT K=2.0, KT=0.0, P1=1.0, P2=0.5
DYNAMIC
   INPUT=STEP(0.0), E=INPUT-Y-KT*Y1DOT, E1=K*E
   Y2DOT=LEADLAG(P1,P2,E1)
   Y1DOT=INTGRL(0.0,Y2DOT)
       Y=INTGRL(0.0,Y1DOT)
  OUTPUT=Y
TERMINAL
  TIMER FINTIME=5.0, PRTDEL=0.05, PRTPLT OUTPUT
END
STOP
END JOB
```

The computer will then print the output for these conditions and, if the results are satisfactory, the problem is solved; if, however the results are poor, different values of KT may be tried. One way of obtaining the step results for these values of added feedback compensation is simply to change KT=0 to several other increasing values and repeat the operation. A better way would be to initially delete KT from the CONSTANT line, then follow with another statement line such as:

```
PARAMETER KT=(0.0, 0.2, 0.4, 0.6, 0.8, 1.0)
```

Still another way would be to do this with a FORTRAN statement along with a NOSORT notation. The FORTRAN statement could be based upon a conditional requirement of some parameter of the output response and then looped until that requirement is achieved; this method depends on the sophistication of the programmer with regard to both FORTRAN and also his knowledge of control system optimal requirements.

DIGITAL COMPUTERS 3

Situation: A particular memory requires a RAM of 1Kx4 bits. Unfortunately only 1Kx1 bits RAMS (say, IC #2125A's)* are available for this particular application. The data from this RAM structure is to be made available along with another identical RAM structure to be added through a group (4) of full adders. The output from these full adders is to be stored for later use in 5 edge triggered D type flip flops.

a) Show the interconnections necessary for the equivalent 1Kx4 bit RAMS (from the 1Kx1 bit RAMS). Call this equivalent circuit RAM1 and another identical unit RAM2.

b) Show the interconnections for the 4 full adders, the RAM's, and the 5 D type flip flops. Assume all data is in the natural binary format.

c) Assume now that data stored in the RAM's is known to be in the binary coded decimal format and that data stored in the D flip flops is to be made available to 2 seven segment drivers and LED displays. Show the interconnections necessary for correcting the full adders to correctly read the BCD sums and the connections to drivers and the LED displays.

*2125A Chip Logic

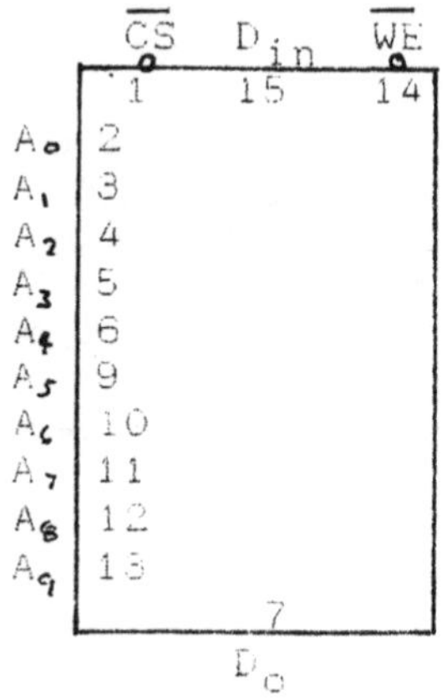

SOLUTION

a)

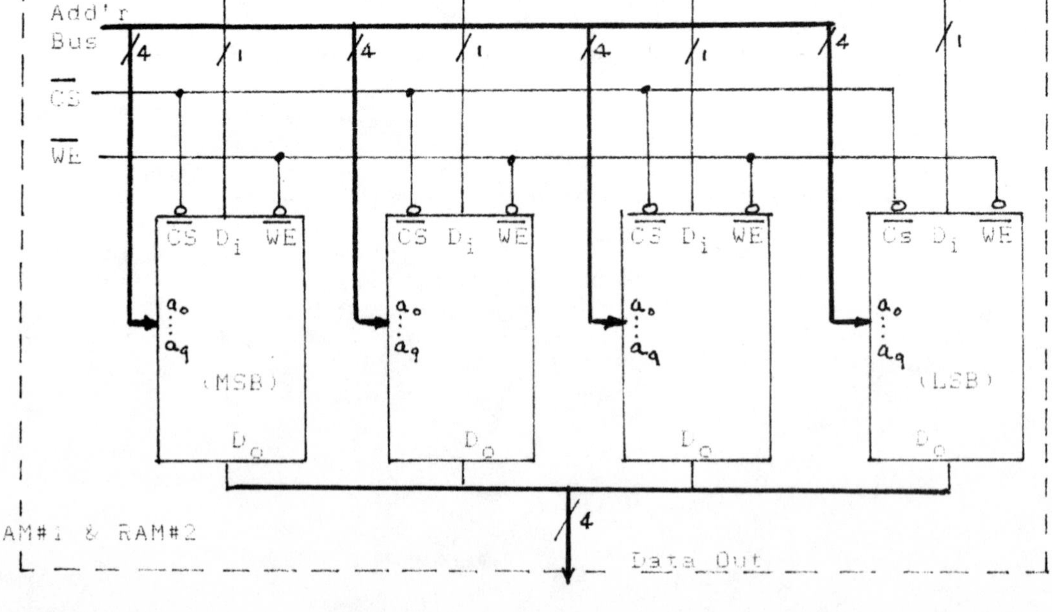

b)

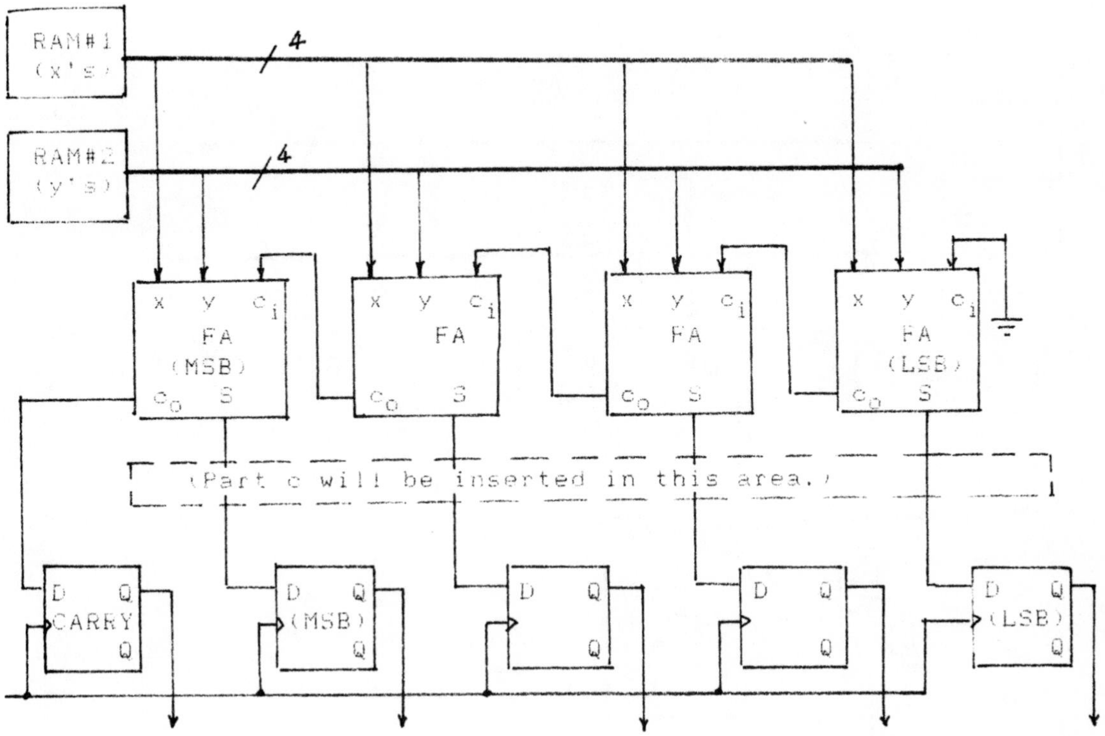

c) To correct the outputs of the natural binary signals to that of binary coded decimal, it must be recalled that natural binary includes numbers between 0 and 15; while BCD numbers are limited to 0 through 9. As long as the sum does not exceed 9 the outputs are correct. However, if the sum exceeds 9, then 6 must be added to the natural binary sum to yield two BCD numbers. As an example, suppose 7 and 5 are added together, the natural binary sum is $(1100)_2$ while if we add 6 more, the new sum is $(0001\ 0010)_{BCD}$.

7 ---->	0111	
5 ---->	0101	
	1100	Sum is greater than 9!
	0110	Therefore add 6.
12---->0001	0010	Answer in BCD.

Thus it is necessary to detect if outputs of S's of the full adder exceed 9, if so add 6 with the additional circuitry. This detection is easily accomplished if the 8 line is high AND either the 4 OR 2 line is high (and, of course, OR if the c_o output of the MSB FA is high).

The two 7 segment driver/decoders are connected only to show a 0 or a 1 for the MSB and a 0-->9 for the LSB.

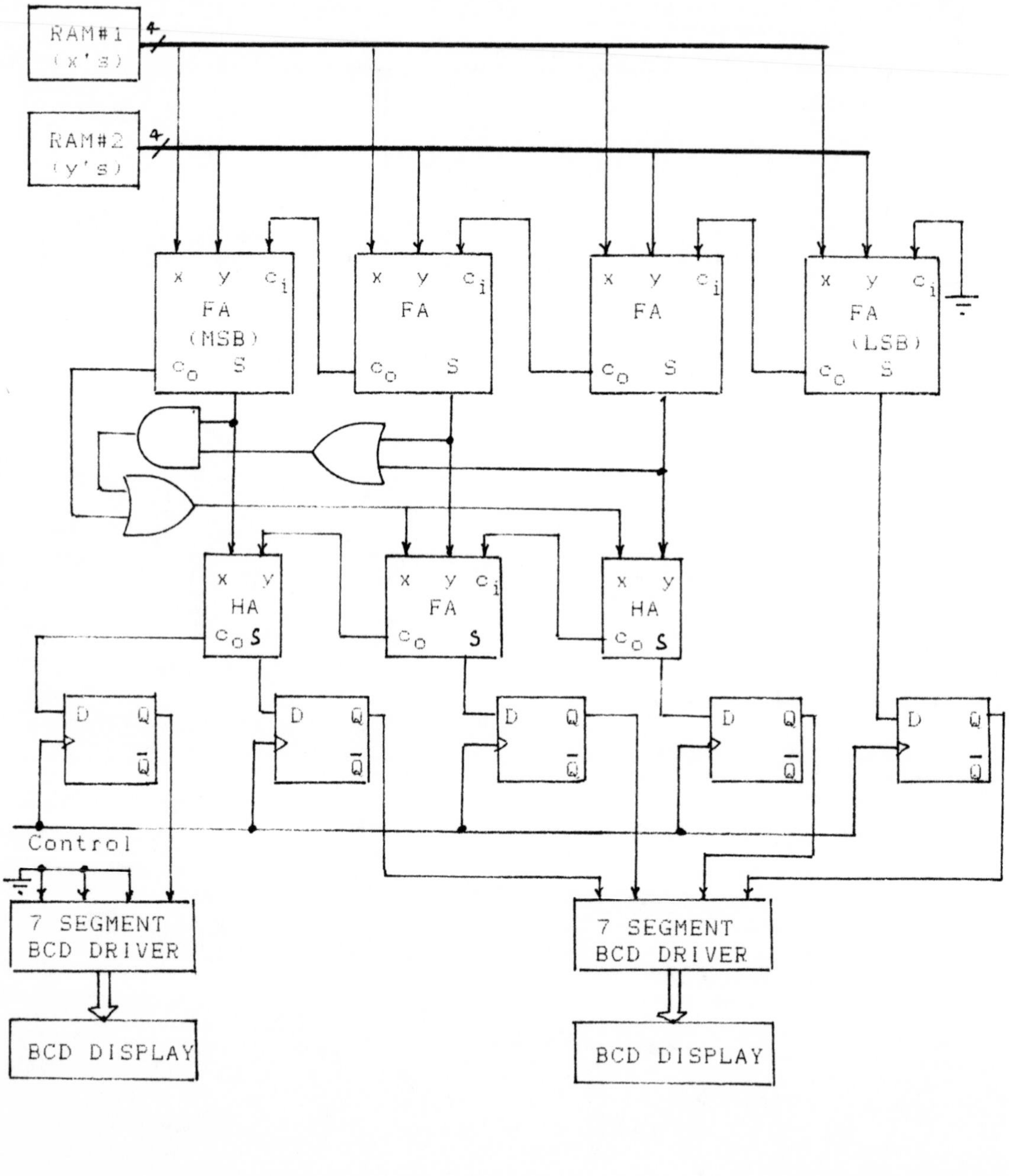
RAM#1
(x's)
4
RAM#2
(y's)
4
x y c_i
FA
(MSB)
c_o S
x y c_i
FA
c_o S
x y c_i
FA
c_o S
x y c_i
FA
(LSB)
c_o S
x y
HA
c_o S
x y c_i
FA
c_o S
x y
HA
c_o S
D Q
$\overline{Q}$
Control
7 SEGMENT
BCD DRIVER
BCD DISPLAY
7 SEGMENT
BCD DRIVER
BCD DISPLAY

DIGITAL COMPUTERS 4

If the lengths of the sides of a triangle are given by the values of the variables X, Y, and Z, then the area of the triangle can be computed from:

$$AREA = \sqrt{W(W-X)(W-Y)(W-Z)} \qquad \text{where} \quad W = \frac{X + Y + Z}{2}$$

You are to write a BASIC computer program to do the following:

- Input values of X, Y, and Z from the keyboard.
- Compute the area of the triangle.
- Output a heading identifying X, Y, Z and AREA, followed by their values in an exponential format.

SOLUTION

```
10 INPUT "INPUT X,Y,Z  ", X,Y,Z
20 W=(X+Y+Z)/2.
30 AREA=(W*(W-X)*(W-Y)*(W-Z))^.5
40 PRINT "SIDE X", "SIDE Y","SIDE Z", "AREA"
50 PRINT USING "#.###^^^^      "; X,Y,Z,AREA
60 END
```

```
LIST
10 INPUT "INPUT X,Y,Z  ",X,Y,Z
20 W=(X+Y+Z)/2.
30 AREA = (W*(W-X)*(W-Y)*(W-Z))^.5
40 PRINT "SIDE X", "SIDE Y", "SIDE Z", "AREA"
50 PRINT USING "#.###^^^^     "; X,Y,Z,AREA
60 END
Ok
```

```
RUN
INPUT X,Y,Z  3.5,14.75,15.10
SIDE X         SIDE Y         SIDE Z         AREA
0.350E+01      0.148E+02      0.151E+02      0.258E+02
Ok
```

DIGITAL COMPUTERS 5

A single sensor has been used to detect excessive levels of a contaminant in a chemical process. Its output is "0" for normal conditions and "1" when the impurity level becomes too high. An evaluation program shows that the sensor is subject to occasional false alarms; also, it sometimes fails to operate when the contaminant level is high.

To improve the situation, it is decided to use three sensors and to disregard the indication of any single sensor whose output differs from the other two sensors. It is desired to have a single "correct signal" output which will be "0" for normal conditions and "1" for an excessive contaminant level, and an "alarm" output which will be "0" when all sensor outputs are identical ("0" or "1") and which will latch into a "1" output at any time when the three sensor outputs become not identical.

Also, an "alarm reset" input is to be provided such than a "1" input to it will reset the alarm output to zero after an alarm condition ceases to exist.

REQUIRED: Derive a logic system to meet the above requirements. The system is to be implemented with NOR logic. There are no "fan in" or "fan out" limitations, i.e. any NOR gate can have as many inputs as may be needed and can drive any necessary number of gates from its output. Your solution should include:

(a) A truth table for the "correct signal" and "alarm" outputs.
(b) The logic equations.
(c) A logic circuit diagram showing how the sensors and NOR gates are to be connected.

SOLUTION

Let sensor outputs be A, B, and C; then a truth table may be formed with f_0 being the contaminant signal and f_A being the alarm.

Term	Sensors			Outputs	
m	A	B	C	f_0	f_A
0	0	0	0	0	0
1	0	0	1	0	1
2	0	1	0	0	1
3	0	1	1	1	1
4	1	0	0	0	1
5	1	0	1	1	1
6	1	1	0	1	1
7	1	1	1	1	0

$$\therefore f_0 = \Sigma m(3,5,6,7)$$

$$\therefore f_A = \Sigma m(1,2,3,4,5,6)$$

Then simplify by use of Karnaugh Map.

For f_0:

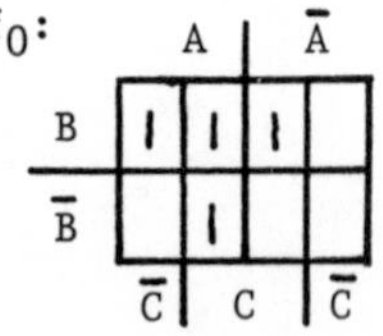

$$\therefore f_0 = AB + BC + AC$$

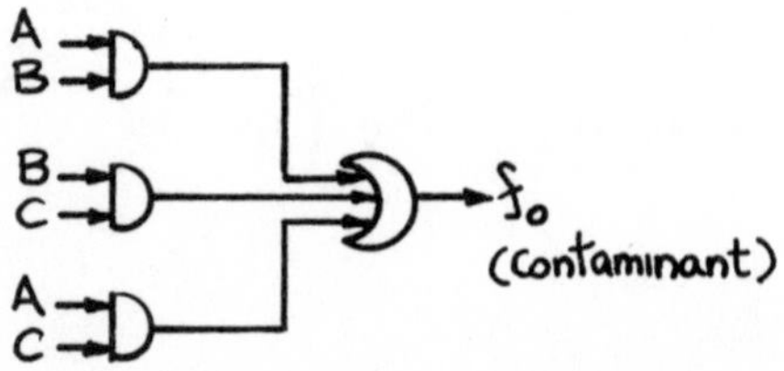

For f_A:

	A		$\bar{A}$	
B	1		1	1
$\bar{B}$	1	1	1	
	$\bar{C}$	C	C	$\bar{C}$

$$\therefore f_A = \bar{A}B + \bar{B}C + A\bar{C}$$

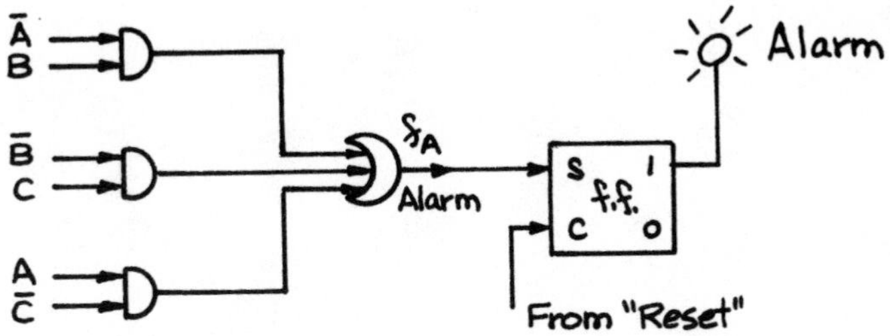

One could have simplified the Karnaugh Map in terms of "O's" as:

$$\bar{f}_o = \bar{B}\bar{C} + \bar{A}\bar{B} + \bar{A}\bar{C}$$

$$\therefore f_o = \overline{\bar{B}\bar{C} + \bar{A}\bar{B} + \bar{A}\bar{C}}$$

$$= (B+C)\cdot(A+C)\cdot(A+B) \text{ to go directly to NOR logic:}$$

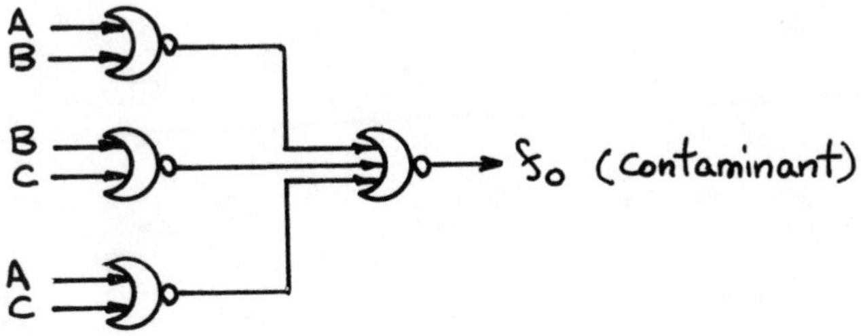

And $\bar{f}_A = ABC + \bar{A}\bar{B}\bar{C}$

$$f_A = \overline{ABC + \bar{A}\bar{B}\bar{C}} = (\bar{A}+\bar{B}+\bar{C})\cdot(A+B+C)$$

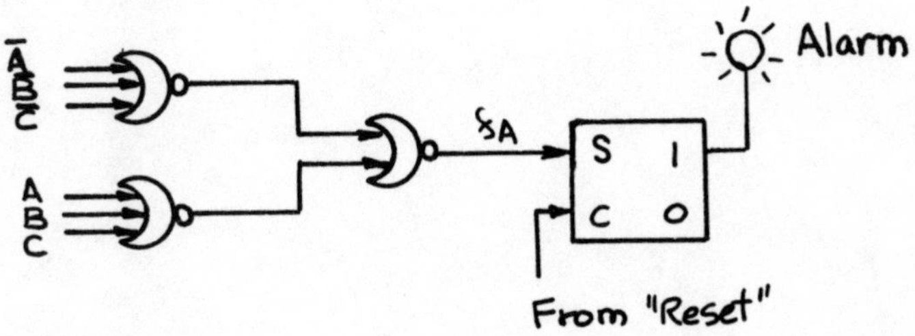

9

Engineering Economics

We begin this chapter with a brief review of engineering economics. The topics cover the full range of what has appeared on past Professional Engineer exams, plus a little more (like geometric gradient and rate of return analysis of multiple alternatives). Engineering economics problems are in the multiple choice format in the P.E. exam.

There are 26 example problems scattered throughout the engineering economics review. These examples are an integral part of the review and should be worked to completion as you come to them.

CASH FLOW

The field of engineering economics uses mathematical and economic techniques to systematically analyze situations which pose alternative courses of action.

The initial step in engineering economics problems is to resolve a situation, or each alternative course in a given situation, into its favorable and unfavorable consequences or factors. These are then measured in some common unit—usually money. Those factors which cannot readily be reduced to money are called intangible, or irreducible, factors. Intangible or irreducible factors are not included in any monetary analysis but are considered in conjunction with such an analysis when making the final decision on proposed courses of action.

A cash flow table shows the "money consequences" of a situation and its timing. For example, a simple problem might be to list the year-by-year consequences of purchasing and owning a used car:

Year	*Cash Flow*	
Beginning of first Year 0	−$4500	Car purchased "now" for $4500 cash. The minus sign indicates a disbursement.
End of Year 1	−350	Maintenance costs are $350 per year.
End of Year 2	−350	
End of Year 3	−350	
End of Year 4	−350	
	+2000	The car is sold at the end of the 4th year for $2000. The plus sign represents a receipt of money.

This same cash flow may be represented graphically:

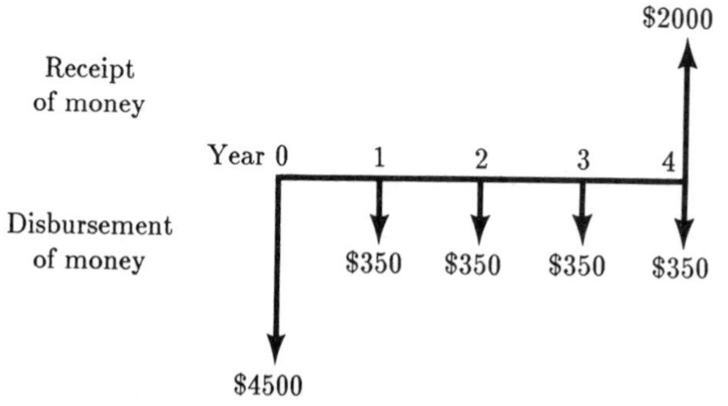

The upward arrow represents a receipt of money, and the downward arrows represent disbursements. The x-axis represents the passage of time.

EXAMPLE 1

In January 1990, a firm purchases a used typewriter for $500. Repairs cost nothing in 1990 or 1991. Repairs are $85 in 1992, $130 in 1993, and $140 in 1994. The machine is sold in 1994 for $300. Compute the cash flow table.

Solution

Unless otherwise stated in problems, the customary assumption is a beginning-of-year purchase, followed by end-of-year receipts or disbursements, and an end-of-year resale or salvage value. Thus the typewriter repairs and the typewriter sale are assumed to occur at the end of the year. Letting a minus sign represent a disbursement of money, and a plus sign a receipt of money, we are able to set up this cash flow table:

Year	*Cash Flow*
Beginning of 1990	−$500
End of 1990	0
End of 1991	0
End of 1992	−85
End of 1993	−130
End of 1994	+160

Notice that at the end of 1994, the cash flow table shows +160 which is the net sum of −140 and +300. If we define Year 0 as the beginning of 1990, the cash flow table becomes:

Year	*Cash Flow*
0	−$500
1	0
2	0
3	−85
4	−130
5	+160

From this cash flow table, the definitions of Year 0 and Year 1 become clear. Year 0 is defined as the *beginning* of Year 1. Year 1 is the *end* of Year 1. Year 2 is the *end* of Year 2, and so forth.

TIME VALUE OF MONEY

When the money consequences of an alternative occur in a short period of time—say, less than one year—we might simply add up the various sums of money and obtain the net result. But we cannot treat money this same way over longer periods of time. This is because money today is not the same as money at some future time.

Consider this question: Which would you prefer, $100 today or the assurance of receiving $100 a year from now? Clearly, you would prefer the $100 today. If you had the money today, rather than a year from now, you could use it for the year. And if you had no use for it, you could lend it to someone who would pay interest for the privilege of using your money for the year.

EQUIVALENCE

In the preceding section we saw that money at different points in time (for example, $100 today or $100 one year hence) may be equal in the sense that they both are $100, but $100 a year hence is *not* an acceptable substitute for $100 today. When we have acceptable substitutes, we say they are *equivalent* to each other. Thus at 8% interest, $108 a year hence is equivalent to $100 today.

EXAMPLE 2
At a 10% per year interest rate, $500 now is *equivalent* to how much three years hence?

Solution
$500 now will increase by 10% in each of the three years.

Now =			$500
End of 1st year =	500 + 10%(500)	=	550
End of 2nd year =	550 + 10%(550)	=	605
End of 3rd year =	605 + 10%(605)	=	665.50

Thus $500 now is *equivalent* to $665.50 at the end of three years.

Equivalence is an essential factor in engineering economic analysis. Suppose we wish to select the better of two alternatives. First, we must compute their cash flows. An example would be:

	Alternative	
Year	*A*	*B*
0	−\$2000	−\$2800
1	+800	+1100
2	+800	+1100
3	+800	+1100

The larger investment in Alternative B results in larger subsequent benefits, but we have no direct way of knowing if Alternative B is better than Alternative A. Therefore we do not know which alternative should be selected. To make a decision we must resolve the alternatives into *equivalent* sums so they may be compared accurately and a decision made.

COMPOUND INTEREST FACTORS

To facilitate equivalence computations a series of compound interest factors will be derived and their use illustrated.

Symbols

i = Interest rate per interest period. In equations the interest rate is stated as a decimal (that is, 8% interest is 0.08).

n = Number of interest periods.

P = A present sum of money.

F = A future sum of money. The future sum F is an amount, n interest periods from the present, that is equivalent to P with interest rate i.

A = An end-of-period cash receipt or disbursement in a uniform series continuing for n periods, the entire series equivalent to P or F at interest rate i.

G = Uniform period-by-period increase in cash flows; the arithmetic gradient.

g = Uniform *rate* of period-by-period increase in cash flows; the geometric gradient.

Functional Notation

	To Find	*Given*	*Functional Notation*
Single Payment			
Compound Amount Factor	F	P	$(F/P,i,n)$
Present Worth Factor	P	F	$(P/F,i,n)$

FUNCTIONAL NOTATION, continued

	To Find	*Given*	*Functional Notation*
Uniform Payment Series			
Sinking Fund Factor	A	F	$(A/F,i,n)$
Capital Recovery Factor	A	P	$(A/P,i,n)$
Compound Amount Factor	F	A	$(F/A,i,n)$
Present Worth Factor	P	A	$(P/A,i,n)$
Arithmetic Gradient			
Gradient Uniform Series	A	G	$(A/G,i,n)$
Gradient Present Worth	P	G	$(P/G,i,n)$

From the table above we can see that the functional notation scheme is based on writing (To Find / Given,i,n). Thus, if we wished to find the future sum F, given a uniform series of receipts A, the proper compound interest factor to use would be $(F/A,i,n)$.

Single Payment Formulas

Suppose a present sum of money P is invested for one year at interest rate i. At the end of the year, we receive back our initial investment P together with interest equal to Pi or a total amount $P + Pi$. Factoring P, the sum at the end of one year is $P(1 + i)$. If we agree to let our investment remain for subsequent years, the progression is as follows:

	Amount at Beginning of Period	+	Interest for the Period	=	Amount at End of the Period
1st year	P	+	Pi	=	$P(1 + i)$
2nd year	$P(1 + i)$	+	$Pi(1 + i)$	=	$P(1 + i)^2$
3rd year	$P(1 + i)^2$	+	$Pi(1 + i)^2$	=	$P(1 + i)^3$
nth year	$P(1 + i)^{n-1}$	+	$Pi(1 + i)^{n-1}$	=	$P(1 + i)^n$

The present sum P increaases in n periods to $P(1 + i)^n$. This gives us a relationship between a present sum P and its equivalent future sum F:

$$\text{Future Sum} = (\text{Present Sum})(1 + i)^n$$
$$F = P(+ i)^n$$

This is the Single Payment Compound Amount formula. In functional notation it is written:

$$F = P(F/P,i,n)$$

The relationship may be rewritten as:

$$\text{Present Sum} = (\text{Future Sum})(1 + i)^{-n}$$
$$P = F(1 + i)^{-n}$$

This is the Single Payment Present Worth formula. It is written:

$$P = F(P/F,i,n)$$

EXAMPLE 3

At a 10% per year interest rate, \$500 now is *equivalent* to how much three years hence?

Solution

This problem was solved in Example 2. Now it can be solved using a single payment formula.

$P =$ \$500
$n =$ 3 years
$i =$ 10%
$F =$ unknown

$$F = P(1 + i)^n = 500(1 + 0.10)^3 = \$665.50$$

This problem may also be solved using the Compound Interest Tables.

$$F = P(F/P,i,n) = 500(F/P,10\%,3)$$

From the 10% Compound Interest Table, read $(F/P,10\%,3) = 1.331$.

$$F = 500(F/P,10\%,3) = 500(1.331) = \$665.50$$

EXAMPLE 4

To raise money for a new business, a man asks you to loan him some money. He offers to pay you \$3000 at the end of four years. How much should you give him now if you want 12% interest per year on your money?

Solution

$P =$ unknown
$n =$ 4 years
$i =$ 12%
$F =$ \$3000

$$P = F(1 + i)^{-n} = 3000(1 + 0.12)^{-4} = \$1906.55$$

Alternate computation using Compound Interest Tables:

$$P = F(P/F,i,n) = 3000(P/F,12\%,4)$$
$$= 3000(0.6355) = \$1906.50$$

Note that the solution based on the Compound Interest Table is slightly different from the exact solution using a hand calculator. In economic analysis, the Compound Interest Tables are always considered to be sufficiently accurate.

Uniform Payment Series Formulas

Consider the following situation:

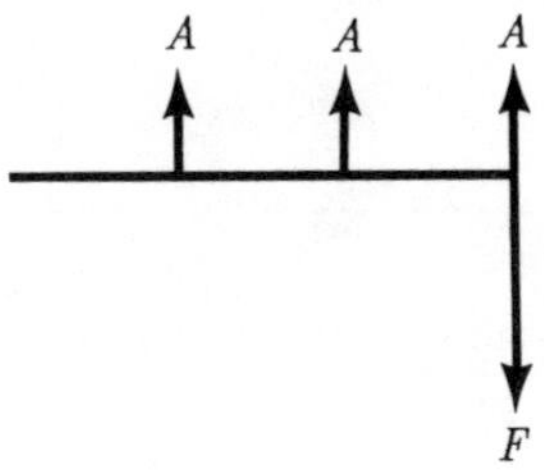

$A =$ End-of-period cash receipt or disbursement in a uniform series continuing for n periods.
$F =$ A future sum of money.

Using the single payment compound amount factor, we can write an equation for F in terms of A:

$$F = A + A(1 + i) + A(1 + i)^2 \tag{1}$$

In our situation, with $n = 3$, Equation (1) may be written in a more general form:

$$F = A + A(1 + i) + A(1 + i)^{n-1} \tag{2}$$

Multiply Eq. (2) by $(1 + i)$: $(1 + i)F = A(1 + i) + A(1 + i)^{n-1} + A(1 + i)^n$ (3)

Write Eq. (2): $- \quad F = A + A(1 + i) + A(1 + i)^{n-1}$ (2)

(3) − (2): $iF = -A + A(1 + i)^n$

$$F = A\left(\frac{(1 + i)^n - 1}{i}\right)$$

Uniform Series Compound Amount formula

Solving this equation for A:

$$A = F\left(\frac{i}{(1 + i)^n - 1}\right)$$ Uniform Series Sinking Fund formula

Since $F = P(1 + i)^n$, we can substitute this expression for F in the equation and obtain:

$$A = P\left(\frac{i(1 + i)^n}{(1 + i)^n - 1}\right)$$ Uniform Series Capital Recovery formula

Solving the equation for P:

$$P = A\left(\frac{(1 + i)^n - 1}{i(1 + i)^n}\right)$$ Uniform Series Present Worth formula

In functional notation, the uniform series factors are:

Compound Amount	$(F/A,i,n)$
Sinking Fund	$(A/F,i,n)$
Capital Recovery	$(A/P,i,n)$
Present Worth	$(P/A,i,n)$

EXAMPLE 5
If \$100 is deposited at the end of each year in a savings account that pays 6% interest per year, how much will be in the account at the end of five years?

Solution

A = \$100
F = unknown
n = 5 years
i = 6%

$$F = A(F/A,i,n) = 100(F/A,6\%,5)$$
$$= 100(5.637) = \$563.70$$

EXAMPLE 6
A woman wishes to make a uniform deposit every three months to her savings account so that at the end of 10 years she will have \$10,000 in the account. If the account earns 6% annual interest, compounded quarterly, how much should she deposit each three months?

Solution

F = \$10,000
A = unknown
n = 40 quarterly deposits
i = $1\frac{1}{2}$% per quarter year

Note that i, the interest rate per interest period, is $1\frac{1}{2}$%, and there are 40 deposits.

$$A = F(A/F,i,n) = 10{,}000(A/F,1\tfrac{1}{2}\%,40)$$
$$= 10{,}000(0.0184) = \$184$$

EXAMPLE 7
An individual is considering the purchase of a used automobile. The total price is \$6200 with \$1240 as a downpayment and the balance paid in 48 equal monthly payments with interest at 1% per month. The payments are due at the end of each month. Compute the monthly payment.

Solution
The amount to be repaid by the 48 monthly payments is the cost of the automobile *minus* the \$1240 downpayment.

$P = \$4960$
$A =$ unknown
$n =$ 48 monthly payments
$i =$ 1% per month

$$A = P(A/P,i,n) = 4960(A/P,1\%,48)$$
$$= 4960(0.0263) = \$130.45$$

EXAMPLE 8

A couple sold their home. In addition to cash, they took a mortgage on the house. The mortgage will be paid off by monthly payments of \$232.50 for 10 years. The couple decides to sell the mortgage to a local bank. The bank will buy the mortgage, but requires a 1% per month interest rate on their investment. How much will the bank pay for the mortgage?

Solution

$A = \$232.50$
$n =$ 120 months
$i =$ 1% per month
$P =$ unknown

$$P = A(P/A,i,n) = 232.50(P/A,1\%,120)$$
$$= 232.50(69.701) = \$16,205.48$$

Arithmetic Gradient

At times one will encounter a situation where the cash flow series is not a constant amount A. Instead it is an increasing series like:

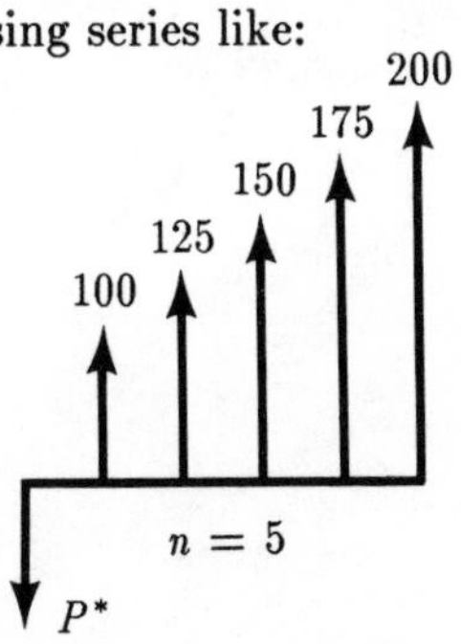

This cash flow may be resolved into two components:

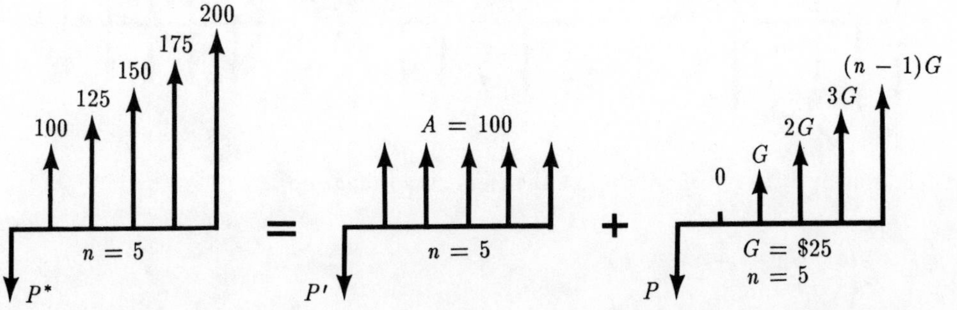

We can compute the value of P^* as equal to P' plus P. We already have an equation for P':

$$P' = A(P/A,i,n)$$

The value for P in the right-hand diagram is:

$$P = G\left(\frac{(1+i)^n - in - 1}{i^2(1+i)^n}\right)$$

This is the Arithmetic Gradient Present Worth formula. In functional notation, the relationship is $P = G(P/G,i,n)$.

EXAMPLE 9

The maintenance on a machine is expected to be \$155 at the end of the first year, and increasing \$35 each year for the following seven years. What present sum of money would need to be set aside now to pay the maintenance for the eight-year period? Assume 6% interest.

Solution

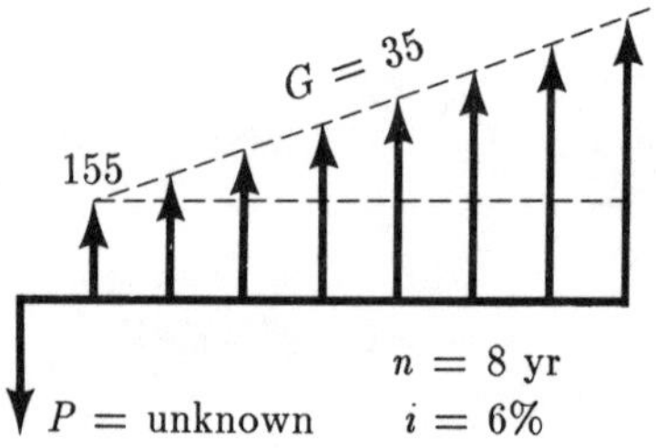

$$\begin{aligned} P &= 155(P/A,6\%,8) + 35(P/G,6\%,8) \\ &= 155(6.210) + 35(19.841) = \$1656.99 \end{aligned}$$

In the gradient series, if instead of the present sum P, an equivalent uniform series A is desired, the problem becomes:

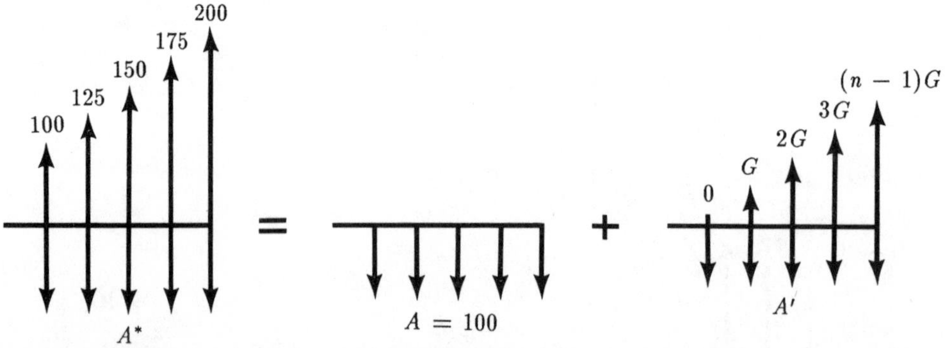

The relationship between A' and G in the right-hand diagram is:

$$A' = G\left(\frac{(1+i)^n - in - 1}{i(1+i)^n - i}\right)$$

In functional notation, the Arithmetic Gradient (to) Uniform Series factor is:

$$A = G(A/G,i,n)$$

It is important to note carefully the diagrams for the two arithmetic gradient series factors. In both cases the first term in the arithmetic gradient series is zero and the last term is $(n - 1)G$. But we use n in the equations and functional notation. The derivations (not shown here) were done on this basis and the arithmetic gradient series Compound Interest Tables are computed this way.

EXAMPLE 10

For the situation in Example 9, we wish now to know the uniform annual maintenance cost. Compute an equivalent A for the maintenance costs.

Solution

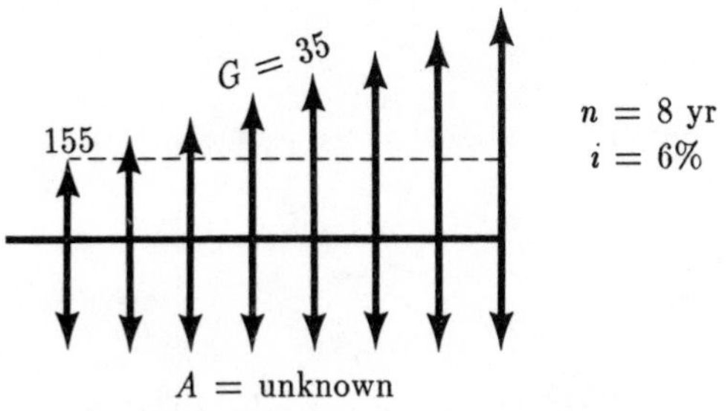

Equivalent uniform annual maintenance cost:

$$\begin{aligned} A &= 155 + 35(A/G,6\%,8) \\ &= 155 + 35(3.195) = \$266.83 \end{aligned}$$

Geometric Gradient

The arithmetic gradient is applicable where the period-by-period change in the cash flow is a uniform amount. There are other situations where the period-by-period change is a *uniform rate, g*. A diagram of this situation is:

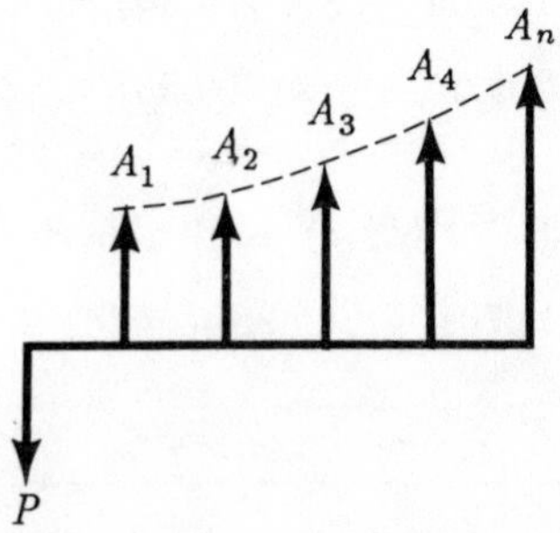

where $A_n = A_1(1 + g)^{n-1}$

g = Uniform rate of period-by-period change; the geometric gradient stated as a decimal (8% = 0.08).

A_1 = Value of A at Year 1.

A_n = Value of A at any Year n.

Geometric Series Present Worth Formulas:

When $i = g$, $P = A_1\left(n(1 + i)^{-1}\right)$

When $i \neq g$, $P = A_1\left(\dfrac{1 - (1 + g)^n(1 + i)^{-n}}{i - g}\right)$

EXAMPLE 11

It is likely that airplane tickets will increase 8% in each of the next four years. The cost of a plane ticket at the end of the first year will be \$180. How much money would need to be placed in a savings account now to have money to pay a student's travel home at the end of each year for the next four years? Assume the savings account pays 5% annual interest.

Solution

The problem describes a geometric gradient where $g = 8\%$ and $i = 5\%$.

$$P = A_1\left(\frac{1 - (1 + g)^n(1 + i)^{-n}}{i - g}\right)$$

$$= 180.00\left(\frac{1 - (1.08)^4(1.05)^{-4}}{0.05 - 0.08}\right) = 180.00\left(\frac{-0.119278}{-0.03}\right) = \$715.67$$

Thus, \$715.67 would need to be deposited now.

As a check, the problem can be solved without using the geometric gradient:

Year		*Ticket*
1	$A_1 =$	\$180.00
2	$A_2 = 180.00 + 8\%(180.00) =$	194.40
3	$A_3 = 194.40 + 8\%(194.40) =$	209.95
4	$A_4 = 209.95 + 8\%(209.95) =$	226.75

$$P = 180.00(P/F,5\%,1) + 194.40(P/F,5\%,2) + 209.95(P/F,5\%,3) + 226.75(P/F,5\%,4)$$

$$= 180.00(0.9524) + 194.40(0.9070) + 209.95(0.8638) + 226.75(0.8227)$$

$$= \$715.66$$

NOMINAL AND EFFECTIVE INTEREST

Nominal interest is the annual interest rate without considering the effect of any compounding.

Effective interest is the annual interest rate taking into account the effect of any compounding during the year.

Frequently an interest rate is described as an annual rate, even though the interest period may be something other than one year. A bank may pay 1-1/2% interest on the amount in a savings account every three months. The *nominal* interest rate in this situation is 6% (4 x 1-1/2% = 6%). But if you deposited \$1000 in such an account, would you have 106%(1000) = \$1060 in the account at the end of one year? The answer is no, you would have more. The amount in the account would increase as follows:

		Amount in Account
	At beginning of year =	\$1000.00
End of 3 months:	$1000.00 + 1\frac{1}{2}\%(1000.00) =$	1015.00
End of 6 months:	$1015.00 + 1\frac{1}{2}\%(1015.00) =$	1030.23
End of 9 months:	$1030.23 + 1\frac{1}{2}\%(1030.23) =$	1045.68
End of one year:	$1045.68 + 1\frac{1}{2}\%(1045.68) =$	1061.37

The actual interest rate on the \$1000 would be the interest, \$61.37, divided by the original \$1000, or 6.137%. We call this the *effective* interest rate.

$$\text{Effective interest rate} = (1 + i)^m - 1, \text{ where}$$

i = Interest rate per interest period;
m = Number of compoundings per year.

EXAMPLE 12
A bank charges $1\frac{1}{2}\%$ per month on the unpaid balance for purchases made on its credit card. What nominal interest rate is it charging? What effective interest rate?

Solution
The nominal interest rate is simply the annual interest ignoring compounding, or $12(1\frac{1}{2}\%) = 18\%$.

$$\text{Effective interest rate} = (1 + 0.015)^{12} - 1 = 0.1956 = 19.56\%$$

SOLVING ECONOMIC ANALYSIS PROBLEMS

The techniques presented so far illustrate how to convert single amounts of money, and uniform or gradient series of money, into some equivalent sum at another point in time. These compound interest computations are an essential part of economic analysis problems.

The typical situation is that we have a number of alternatives and the question is, which alternative should be selected? The customary method of solution is to resolve each of the alternatives into some common form and then choose the best alternative (taking both the monetary and intangible factors into account).

Criteria

Economic analysis problems inevitably fall into one of three categories:

1. Fixed Input — The amount of money or other input resources is fixed.

   Example: A project engineer has a budget of $450,000 to overhaul a plant.

2. Fixed Output — There is a fixed task, or other output to be accomplished.

   Example: A mechanical contractor has been awarded a fixed price contract to air-condition a building.

3. Neither Input nor Output Fixed — This is the general situation where neither the amount of money or other inputs, nor the amount of benefits or other outputs are fixed.

   Example: A consulting engineering firm has more work available than it can handle. It is considering paying the staff for working evenings to increase the amount of design work it can perform.

There are five major methods of comparing alternatives: present worth; future worth; annual cost; rate of return; and benefit–cost ratio. These are presented in the following sections.

PRESENT WORTH

In present worth analysis, the approach is to resolve all the money consequences of an alternative into an equivalent present sum. For the three categories given above, the criteria are:

Category	*Present Worth Criterion*
Fixed Input	Maximize the Present Worth of benefits or other outputs.
Fixed Output	Minimize the Present Worth of costs or other inputs.
Neither Input nor Output Fixed	Maximize [Present Worth of benefits *minus* Present Worth of costs] or, stated another way: Maximize Net Present Worth.

Application of Present Worth

Present worth analysis is most frequently used to determine the present value of future money receipts and disbursements. We might want to know, for example, the present worth of an income producing property, like an oil well. This should provide an estimate of the price at which the property could be bought or sold.

An important restriction in the use of present worth calculations is that there must be a common analysis period when comparing alternatives. It would be incorrect, for example, to compare the present worth (PW) of cost of Pump A, expected to last 6 years, with the PW of cost of Pump B, expected to last 12 years.

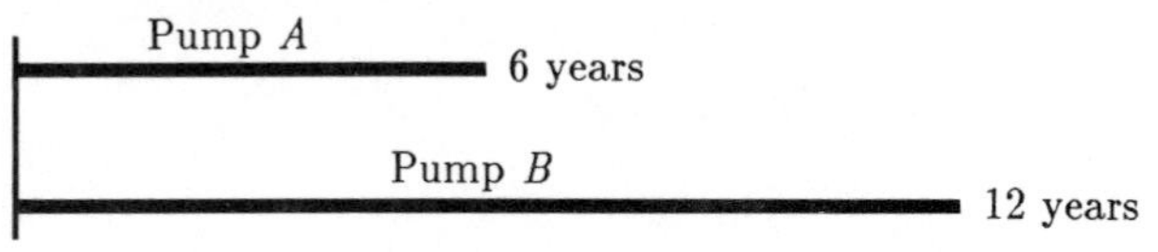

Improper Present Worth Comparison

In situations like this, the solution is either to use some other analysis technique* or to restructure the problem so there is a common analysis period. In the example above, a customary assumption would be that a pump is needed for 12 years and that Pump A will be replaced by an identical Pump A at the end of 6 years. This gives a 12-year common analysis period.

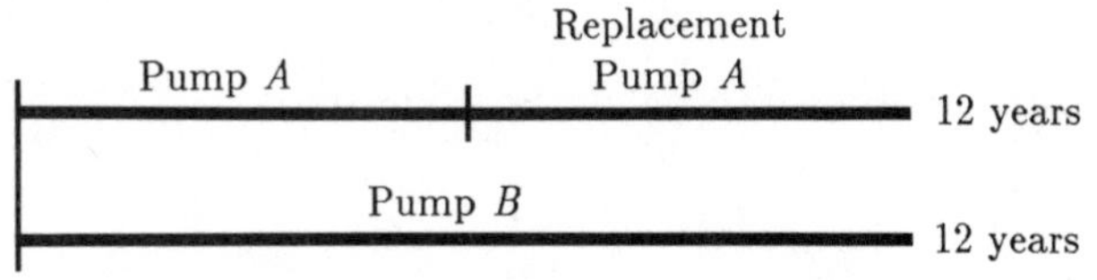

Correct Present Worth Comparison

This approach is easy to use when the different lives of the alternatives have a practical least common multiple life. When this is not true (for example, life of J equals 7 years and the life of K equals 11 years), some assumptions must be made to select a suitable common analysis period, or the present worth method should not be used.

EXAMPLE 13

Machine X has an initial cost of $10,000, annual maintenance of $500 per year, and no salvage value at the end of its four-year useful life. Machine Y costs $20,000. The first year there is no maintenance cost. The second year, maintenance is $100, and increases $100 per year in subsequent years. The machine has an anticipated $5000 salvage value at the end of its 12-year useful life.

If interest is 8%, which machine should be selected?

Solution

The analysis period is not stated in the problem. Therefore we select the least common multiple of the lives, or 12 years, as the analysis period.

*Generally the annual cost method is suitable in these situations.

Present Worth of Cost of 12 years of Machine X

$$\begin{aligned} &= 10{,}000 + 10{,}000(P/F,8\%,4) + 10{,}000(P/F,8\%,8) + 500(P/A,8\%,12) \\ &= 10{,}000 + 10{,}000(0.7350) + 10{,}000(0.5403) + 500(7.536) \\ &= \$26{,}521 \end{aligned}$$

Present Worth of Cost of 12 years of Machine Y

$$\begin{aligned} &= 20{,}000 + 100(P/G,8\%,12) - 5000(P/F,8\%,12) \\ &= 20{,}000 + 100(34.634) - 5000(0.3971) \\ &= \$21{,}478 \end{aligned}$$

Choose Machine Y with its smaller PW of Cost.

EXAMPLE 14
Two alternatives have the following cash flows:

	Alternative	
Year	*A*	*B*
0	−$2000	−$2800
1	+800	+1100
2	+800	+1100
3	+800	+1100

At a 5% interest rate, which alternative should be selected?

Solution
Solving by Present Worth analysis:

Net Present Worth (NPW) = PW of benefits − PW of cost

$$\begin{aligned} \text{NPW}_A &= 800(P/A,5\%,3) - 2000 \\ &= 800(2.723) - 2000 \\ &= +178.40 \end{aligned}$$

$$\begin{aligned} \text{NPW}_B &= 1100(P/A,5\%,3) - 2800 \\ &= 1100(2.723) - 2800 \\ &= +195.30 \end{aligned}$$

To maximize NPW, choose Alternative B.

Capitalized Cost

In the special situation where the analysis period is infinite ($n = \infty$), an analysis of the present worth of cost is called *capitalized cost*. There are a few public projects where the analysis period is infinity. Other examples would be permanent endowments and cemetery perpetual care.

When n equals infinity, a present sum P will accrue interest of Pi for every future interest period. For the principal sum P to continue undiminished (an essential requirement for n equal to infinity), the end-of-period sum A that can be disbursed is Pi.

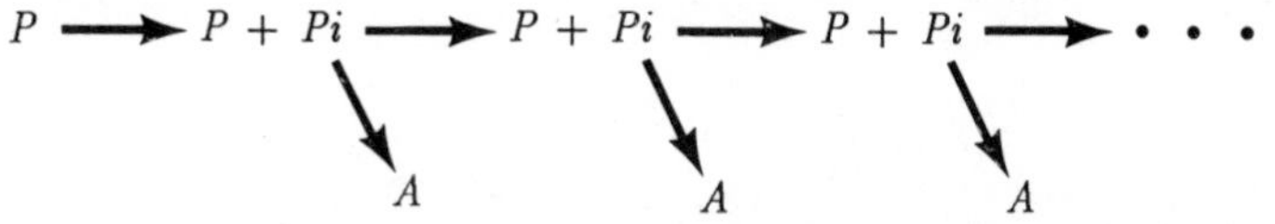

When $n = \infty$, the fundamental relationship between P, A, and i is:

$$A = Pi$$

Some form of this equation is used whenever there is a problem with an infinite analysis period.

EXAMPLE 15

In his will, a man wishes to establish a perpetual trust to provide for the maintenance of a small local park. If the annual maintenance is $7500 per year and the trust account can earn 5% interest, how much money must be set aside in the trust?

Solution

When $n = \infty$, $A = Pi$ or $P = \frac{A}{i}$

$$\text{Capitalized cost } P = \frac{A}{i} = \frac{\$7500}{0.05} = \$150{,}000$$

FUTURE WORTH

In present worth analysis, the comparison is made in terms of the equivalent present costs and benefits. But the analysis need not be made at the present time, it could be made at any point in time: past, present, or future. Although the numerical calculations may look different, the decision is unaffected by the point in time selected. Of course, there are situations where we do want to know what the future situation will be if we take some particular course of action now. When an analysis is made based on some future point in time, it is called future worth analysis.

Category	*Future Worth Criterion*
Fixed Input	Maximize the Future Worth of benefits or other outputs.
Fixed Output	Minimize the Future Worth of costs or other inputs.
Neither Input nor Output Fixed	Maximize [Future Worth of benefits *minus* Future Worth of costs] or, stated another way: Maximize Net Future Worth.

EXAMPLE 16
Two alternatives have the following cash flows:

	Alternative	
Year	*A*	*B*
0	−\$2000	−\$2800
1	+800	+1100
2	+800	+1100
3	+800	+1100

At a 5% interest rate, which alternative should be selected?

Solution
In Example 14, this problem was solved by Present Worth analysis at Year 0. Here it will be solved by Future Worth analysis at the end of Year 3.

$$\text{Net Future Worth (NFW)} = \text{FW of benefits} - \text{FW of cost}$$

$$\begin{aligned} \text{NFW}_A &= 800(F/A,5\%,3) - 2000(F/P,5\%,3) \\ &= 800(3.152) - 2000(1.158) \\ &= +205.60 \end{aligned}$$

$$\begin{aligned} \text{NFW}_B &= 1100(F/A,5\%,3) - 2800(F/P,5\%,3) \\ &= 1100(3.152) - 2800(1.158) \\ &= +224.80 \end{aligned}$$

To maximize NFW, choose Alternative *B*.

ANNUAL COST

The annual cost method is more accurately described as the method of Equivalent Uniform Annual Cost (EUAC) or, where the computation is of benefits, the method of Equivalent Uniform Annual Benefits (EUAB).

Criteria

For each of the three possible categories of problems, there is an annual cost criterion for economic efficiency.

Category	*Annual Cost Criterion*
Fixed Input	Maximize the Equivalent Uniform Annual Benefits. That is, maximize EUAB.
Fixed Output	Minimize the Equivalent Uniform Annual Cost. That is, minimize EUAC.
Neither Input nor Output Fixed	Maximize [EUAB − EUAC].

Application of Annual Cost Analysis

In the section on present worth, we pointed out that the present worth method requires that there be a common analysis period for all alternatives. This same restriction does not apply in all annual cost calculations, but it is important to understand the circumstances that justify comparing alternatives with different service lives.

Frequently an analysis is to provide for a more or less continuing requirement. One might need to pump water from a well, for example, as a continuing requirement. Regardless of whether the pump has a useful service life of 6 years or 12 years, we would select the one whose annual cost is a minimum. And this would still be the case if the pump useful lives were the more troublesome 7 and 11 years, respectively. Thus, if we can assume a continuing need for an item, an annual cost comparison among alternatives of differing service lives is valid.

The underlying assumption made in these situations is that when the shorter-lived alternative has reached the end of its useful life, it can be replaced with an identical item with identical costs, and so forth. This means the EUAC of the initial alternative is equal to the EUAC for the continuing series of replacements.

If, on the other hand, there is a specific requirement in some situation to pump water for 10 years, then each pump must be evaluated to see what costs will be incurred during the analysis period and what salvage value, if any, may be recovered at the end of the analysis period. The annual cost comparison needs to consider the actual circumstances of the situation.

Examination problems are often readily solved by the annual cost method. And the underlying "continuing requirement" is often present, so that an annual cost comparison of unequal-lived alternatives is an appropriate method of analysis.

EXAMPLE 17

Consider the following alternatives:

	A	*B*
First cost	\$5000	\$10,000
Annual maintenance	500	200
End-of-useful-life salvage value	600	1000
Useful life	5 years	15 years

Based on an 8% interest rate, which alternative should be selected?

Solution

Assuming both alternatives perform the same task and there is a continuing requirement, the goal is to minimize EUAC.

Alternative A:

$$\text{EUAC} = 5000(A/P,8\%,5) + 500 - 600(A/F,8\%,5)$$

$$= 5000(0.2505) + 500 - 600(0.1705) = \$1650$$

Alternative B:

$$\text{EUAC} = 10{,}000(A/P{,}8\%{,}15) + 200 - 1000(A/F{,}8\%{,}15)$$
$$= 10{,}000(0.1168) + 200 - 1000(0.0368) = \$1331$$

To minimize EUAC, select Alternative *B*.

RATE OF RETURN

A typical situation is a cash flow representing the costs and benefits. The rate of return may be defined as the interest rate where

PW of cost = PW of benefits,

EUAC = EUAB,

or PW of cost − PW of benefits = 0.

EXAMPLE 18

Compute the rate of return for the investment represented by the following cash flow:

Year	*Cash Flow*
0	−$595
1	+250
2	+200
3	+150
4	+100
5	+50

Solution

This declining arithmetic gradient series may be separated into two cash flows for which compound interest factors are available:

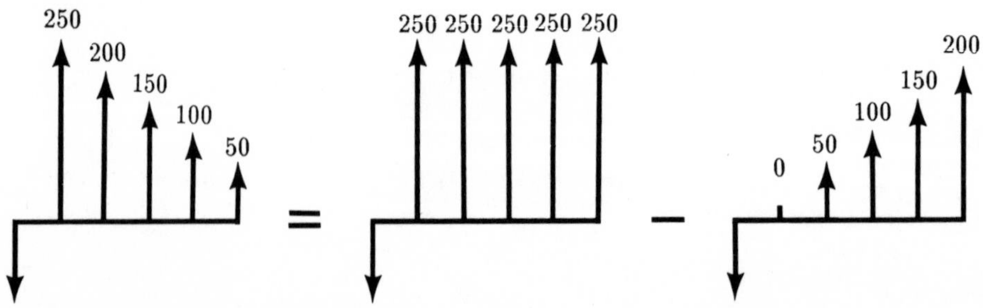

Note that the gradient series factors are based on an *increasing* gradient. Here, the declining cash flow is solved by subtracting an increasing arithmetic gradient, as indicated by the diagram.

PW of cost − PW of benefits = 0

$595 - [250(P/A,i,5) - 50(P/G,i,5)] = 0$

Try $i = 10\%$:

$595 - [250(3.791) - 50(6.862)] = -9.65$

Try $i = 12\%$:

$595 - [250(3.605) - 50(6.397)] = +13.60$

The rate of return is between 10% and 12%. It may be computed more accurately by linear interpolation:

$$\text{Rate of return} = 10\% + (2\%)\left(\frac{9.65 - 0}{13.60 + 9.65}\right) = 10.83\%$$

Rate of Return Criterion for Two Alternatives

Compute the incremental rate of return on the cash flow representing the difference between the two alternatives. Since we want to look at increments of *investment*, the cash flow for the difference between the alternatives is computed by taking the higher initial-cost alternative *minus* the lower initial-cost alternative. If the incremental rate of return is greater than or equal to the predetermined minimum attractive rate of return (MARR), choose the higher-cost alternative; otherwise, choose the lower-cost alternative.

EXAMPLE 19

Two alternatives have the following cash flows:

	Alternative	
Year	*A*	*B*
0	−\$2000	−\$2800
1	+800	+1100
2	+800	+1100
3	+800	+1100

If 5% is considered the minimum attractive rate of return (MARR), which alternative should be selected?

Solution

These two alternatives were previously examined in Examples 14 and 16 by present worth and future worth analysis. This time, the alternatives will be resolved using rate of return analysis.

Note that the problem statement specifies a 5% minimum attractive rate of raturn (MARR), while Examples 14 and 16 referred to a 5% interest rate. These are really two different ways of saying the same thing: the minimum acceptable time value of money is 5%.

First, tabulate the cash flow that represents the increment of investment between the alternatives. This is done by taking the higher initial-cost alternative minus the lower initial-cost alternative:

Year	Alternative A	Alternative B	Difference between Alternatives B − A
0	−$2000	−$2800	−$800
1	+800	+1100	+300
2	+800	+1100	+300
3	+800	+1100	+300

Then compute the rate of return on the increment of investment represented by the difference between the alternatives:

$$\text{PW of cost} = \text{PW of benefits}$$
$$800 = 300(P/A,i,3)$$
$$(P/A,i,3) = \frac{800}{300} = 2.67$$
$$i \approx 6.1\%$$

Since the incremental rate of return exceeds the 5% MARR, the increment of investment is desirable. Choose the higher-cost Alternative *B*.

Before leaving this example problem, one should note something that relates to the rates of return on Alternative *A* and on Alternative *B*. These rates of return, if computed, are:

	Rate of Return
Alternative *A*	9.7%
Alternative *B*	8.7%

The correct answer to this problem has been shown to be Alternative *B*, and this is true even though Alternative *A* has a higher rate of return. The higher-cost alternative may be thought of as the lower-cost alternative, plus the increment of investment between them. Looked at this way, the higher-cost Alternative *B* is equal to the desirable lower-cost Alternative *A* plus the desirable differences between the alternatives.

The important conclusion is that computing the rate of return for each alternative does not provide the basis for choosing between alternatives. Instead, incremental analysis is required.

EXAMPLE 20

Consider the following:

Year	Alternative A	Alternative B
0	−$200.0	−$131.0
1	+77.6	+48.1
2	+77.6	+48.1
3	+77.6	+48.1

If the minimum attractive rate of return (MARR) is 10%, which alternative should be selected?

Solution
To examine the increment of investment between the alternatives, we will examine the higher initial-cost alternative minus the lower initial-cost alternative, or $A - B$.

	Alternative		*Increment*
Year	*A*	*B*	*A − B*
0	−$200.0	−$131.0	−$69.0
1	+77.6	+48.1	+29.5
2	+77.6	+48.1	+29.5
3	+77.6	+48.1	+29.5

Solve for the incremental rate of return:

$$\text{PW of cost} = \text{PW of benefits}$$
$$69.0 = 29.5(P/A,i,3)$$
$$(P/A,i,3) = \frac{69.0}{29.5} = 2.339$$

From Compound Interest Tables, the incremental rate of return is between 12% and 15%. This is a desirable increment of investment hence we select the higher initial-cost Alternative A.

Rate of Return Criterion for Three or More Alternatives

When there are three or more mutually exclusive alternatives, one must proceed following the same general logic presented for two alternatives. The components of incremental analysis are:

1. Compute the rate of return for each alternative. Reject any alternative where the rate of return is less than the given MARR. (This step is not essential, but helps to immediately identify unacceptable alternatives.)

2. Rank the remaining alternatives in their order of increasing initial cost.

3. Examine the increment of investment between the two lowest-cost alternatives as described for the two-alternative problem. Select the best of the two alternatives and reject the other one.

4. Take the preferred alternative from Step 3. Consider the next higher initial-cost alternative and proceed with another two-alternative comparison.

5. Continue until all alternatives have been examined and the best of the multiple alternatives has been identified.

EXAMPLE 21

Consider the following:

	Alternative	
Year	*A*	*B*
0	−\$200.0	−\$131.0
1	+77.6	+48.1
2	+77.6	+48.1
3	+77.6	+48.1

If the minimum attractive rate of return (MARR) is 10%, which alternative, if any, should be selected?

Solution

One should carefully note that this is a *three*-alternative problem where the alternatives are *A*, *B*, and "Do Nothing."

In this solution we will skip Step 1. Reorganize the problem by placing the alternatives in order of increasing initial cost:

	Do	*Alternative*	
Year	*Nothing*	*B*	*A*
0	0	−\$131.0	−\$200.0
1	0	+48.1	+77.6
2	0	+48.1	+77.6
3	0	+48.1	+77.6

Examine the "*B* − Do Nothing" increment of investment:

Year	*B* − Do Nothing
0	−\$131.0 − 0 = −\$131.0
1	+48.1 − 0 = +48.1
2	+48.1 − 0 = +48.1
3	+48.1 − 0 = +48.1

Solve for the incremental rate of return:

$$\begin{aligned} \text{PW of cost} &= \text{PW of benefits} \\ 131.0 &= 48.1(P/A,i,3) \\ (P/A,i,3) &= \frac{131.0}{48.1} = 2.723 \end{aligned}$$

From Compound Interest Tables, the incremental rate of return = 5%. Since the incremental rate of return is less than 10%, the *B* − Do Nothing increment is not desirable. Reject Alternative *B*.

Next, consider the increment of investment between the two remaining alternatives:

Year	*A* − Do Nothing
0	−\$200.0 − 0 = −\$200.0
1	+77.6 − 0 = +77.6
2	+77.6 − 0 = +77.6
3	+77.6 − 0 = +77.6

Solve for the incremental rate of return:

$$\text{PW of cost} = \text{PW of benefits}$$
$$200.0 = 77.6(P/A,i,3)$$
$$(P/A,i,3) = \frac{200.0}{77.6} = 2.577$$

The incremental rate of return is 8%. Since the rate of return on the A – Do Nothing increment of investment is less than the desired 10%, reject the increment by rejecting Alternative A. We select the remaining alternative: Do nothing!

If you have not already done so, you should go back to Example 20 and see how the slightly changed wording of the problem radically altered it. Example 20 required the choice between two undesirable alternatives. Example 21 adds the Do-nothing alternative which is superior to A or B.

EXAMPLE 22

Consider four mutually exclusive alternatives:

	Alternative			
	A	*B*	*C*	*D*
Initial Cost	$400.0	$100.0	$200.0	$500.0
Uniform Annual Benefit	100.9	27.7	46.2	125.2

Each alternative has a five-year useful life and no salvage value. If the minimum attractive rate of return (MARR) is 6%, which alternative should be selected?

Solution

Mutually exclusive is where selecting one alternative precludes selecting any of the other alternatives. This is the typical "textbook" situation. The solution will follow the several steps in incremental analysis.

1. The rate of return is computed for the four alternatives.

Alternative	*A*	*B*	*C*	*D*
Computed rate of return	8.3%	11.9%	5%	8%

Since Alternative C has a rate of return less than the MARR, it may be eliminated from further consideration.

2. Rank the remaining alternatives in order of increasing intial cost and examine the increment between the two lowest cost alternatives.

Alternative	*B*	*A*	*D*
Initial Cost	$100.0	$400.0	$500.0
Uniform Annual Benefit	27.7	100.9	125.2

	A – B
Δ Initial Cost	$300.0
Δ Uniform Annual Benefit	73.2
Computed Δ rate of return	7%

Since the incremental rate of return exceeds the 6% MARR, the increment of investment is desirable. Alternative A is the better alternative.

3. Take the preferred alternative from the previous step and consider the next higher-cost alternative. Do another two-alternative comparison.

	$D - A$
Δ Initial Cost	\$100.0
Δ Uniform Annual Benefit	24.3
Computed Δ rate of return	6.9%

The incremental rate of return exceeds MARR, hence the increment is desirable. Alternative D is preferred over Alternative A.

Conclusion: Select Alternative D. Note that once again the alternative with the highest rate of return (Alt. B) is *not* the proper choice.

BENEFIT-COST RATIO

Generally in public works and governmental economic analyses, the dominant method of analysis is called benefit–cost ratio. It is simply the ratio of benefits divided by costs, taking into account the time value of money.

$$\text{B/C} = \frac{\text{PW of benefits}}{\text{PW of cost}} = \frac{\text{Equivalent Uniform Annual Benefits}}{\text{Equivalent Uniform Annual Cost}}$$

For a given interest rate, a B/C ratio ≥ 1 reflects an acceptable project. The method of analysis using B/C ratio is parallel to that of rate of return analysis. The same kind of incremental analysis is required.

B/C Ratio Criterion for Two Alternatives

Compute the incremental B/C ratio for the cash flow representing the increment of investment between the higher initial-cost alternative and the lower initial-cost alternative. If this incremental B/C ratio is ≥ 1, choose the higher-cost alternative; otherwise, choose the lower-cost alternative.

B/C Ratio Criterion for Three or More Alternatives

Follow the logic for rate of return, except that the test is whether or not the incremental B/C ratio is ≥ 1.

EXAMPLE 23

Solve Example 22 using Benefit–Cost ratio analysis. Consider four mutually exclusive alternatives:

	Alternative			
	A	*B*	*C*	*D*
Initial Cost	\$400.0	\$100.0	\$200.0	\$500.0
Uniform Annual Benefit	100.9	27.7	46.2	125.2

Each alternative has a five-year useful life and no salvage value. Based on a 6% interest rate, which alternative should be selected?

Solution

1. B/C ratio computed for the alternatives:

$$\text{Alt. } A \quad \text{B/C} = \frac{\text{PW of benefits}}{\text{PW of cost}} = \frac{100.9(P/A,6\%,5)}{400} = 1.06$$

$$B \quad \text{B/C} = \frac{27.7(P/A,6\%,5)}{100} = 1.17$$

$$C \quad \text{B/C} = \frac{46.2(P/A,6\%,5)}{200} = 0.97$$

$$D \quad \text{B/C} = \frac{125.2(P/A,6\%,5)}{500} = 1.05$$

Alternative C with a B/C ratio less than 1 is eliminated.

2. Rank the remaining alternatives in order of increasing intial cost and examine the increment of investment between the two lowest cost alternatives.

Alternative	*B*	*A*	*D*
Initial Cost	\$100.0	\$400.0	\$500.0
Uniform Annual Benefit	27.7	100.9	125.2

	A − *B*
Initial Cost	\$300.0
Uniform Annual Benefit	73.2

$$\text{Incremental B/C ratio} = \frac{73.2(P/A,6\%,5)}{300} = 1.03$$

The incremental B/C ratio exceeds 1.0 hence the increment is desirable. Alternative A is preferred over B.

3. Do the next two-alternative comparison.

	Alternative		*Increment*
	A	*D*	*D* − *A*
Initial Cost	\$400.0	\$500.0	\$100.0
Uniform Annual Benefit	100.9	125.2	24.3

$$\text{Incremental B/C ratio} = \frac{24.3(P/A,6\%,5)}{100} = 1.02$$

The incremental B/C ratio exceeds 1.0, hence Alternative D is preferred.

Conclusion: Select Alternative D.

BREAKEVEN ANALYSIS

In business, "breakeven" is defined as the point where income just covers the associated costs. In engineering economics, the breakeven point is more precisely defined as the point where two alternatives are equivalent.

EXAMPLE 24

A city is considering a new \$50,000 snowplow. The new machine will operate at a savings of \$600 per day, compared to the equipment presently being used. Assume the minimum attractive rate of return (interest rate) is 12% and the machine's life is 10 years with zero resale value at that time. How many days per year must the machine be used to make the investment economical?

Solution

This breakeven problem may be readily solved by annual cost computations. We will set the equivalent uniform annual cost of the snowplow equal to its annual benefit, and solve for the required annual utilization.

Let X = breakeven point = days of operation per year.

$$\text{EUAC} = \text{EUAB}$$

$$50{,}000(A/P,12\%,10) = 600X$$

$$X = \frac{50{,}000(0.1770)}{600} = 14.7 \text{ days/year}$$

DEPRECIATION

Depreciation of captial equipment is an important component of many after-tax economic analyses. For this reason, one must understand the fundamentals of depreciation accounting.

Depreciation is defined, in its accounting sense, as the systematic allocation of the cost of a capital asset over its useful life. *Book value* is defined as the original cost of an asset, minus the accumulated depreciation of the asset.

In computing a schedule of depreciation charges, three items are considered:

1. Cost of the property, P;
2. Depreciable life in years, n;
3. Salvage value of the property at the end of its depreciable life, S.

Straight Line Depreciation

$$\text{Depreciation charge in any year} = \frac{P - S}{n}$$

Sum-Of-Years-Digits Depreciation

$$\begin{array}{c}\text{Depreciation charge}\\ \text{in any year}\end{array} = \frac{\text{Remaining Depreciable Life at Beginning of Year}}{\text{Sum of Years Digits for Total Useful Life}}(P - S)$$

where Sum Of Years Digits $= 1 + 2 + 3 + \cdots + n = \frac{n}{2}(n + 1)$

Double Declining Balance Depreciation

$$\text{Depreciation charge in any year} = \frac{2}{n}(P - \text{Depreciation charges to date})$$

Accumulated Cost Recovery System (ACRS) Depreciation

ACRS depreciation is based on a property class life which is generally less than the actual useful life of the property and on zero salvage value. The varying depreciation percentage to use must be read from a table (based on declining balance with conversion to straight line). Unless one knows the proper ACRS property class, and has an ACRS depreciation table, the depreciation charge in any year cannot be computed.

EXAMPLE 25

A piece of machinery costs \$5000 and has an anticipated \$1000 salvage value at the end of its five-year depreciable life. Compute the depreciation schedule for the machinery by:

(a) Straight line depreciation;

(b) Sum-of-years-digits depreciation;

(c) Double declining balance depreciation.

Solution

$$\text{Straight line depreciation} = \frac{P - S}{n} = \frac{5000 - 1000}{5} = \$800$$

Sum-of-years-digits depreciation:

$$\text{Sum-of-years-digits} = \frac{n}{2}(n + 1) = \frac{5}{2}(6) = 15$$

$$\text{1st year depreciation} = \frac{5}{15}(5000 - 1000) = \$1333$$

$$\text{2nd year depreciation} = \tfrac{4}{15}(5000 - 1000) = 1067$$

$$\text{3rd year depreciation} = \tfrac{3}{15}(5000 - 1000) = 800$$

$$\text{4th year depreciation} = \tfrac{2}{15}(5000 - 1000) = 533$$

$$\text{5th year depreciation} = \tfrac{1}{15}(5000 - 1000) = \underline{267}$$

$$\$4000$$

Double declining balance depreciation:

$$\text{1st year depreciation} = \tfrac{2}{5}(5000 - 0) = \$2000$$

$$\text{2nd year depreciation} = \tfrac{2}{5}(5000 - 2000) = 1200$$

$$\text{3rd year depreciation} = \tfrac{2}{5}(5000 - 3200) = 720$$

$$\text{4th year depreciation} = \tfrac{2}{5}(5000 - 3920) = 432$$

$$\text{5th year depreciation} = \tfrac{2}{5}(5000 - 4352) = \underline{259}$$

$$\$4611$$

Since the problem specifies a $1000 salvage value, the total depreciation may not exceed $4000. The double declining balance depreciation must be stopped in the 4th year when it totals $4000.

The depreciation schedules computed by the three methods are as follows:

Year	*Straight Line*	*Sum-Of-Years-Digits*	*Double Declining Balance*
1	$800	$1333	$2000
2	800	1067	1200
3	800	800	720
4	800	533	80
5	800	267	0
	$4000	$4000	$4000

INCOME TAXES

Income taxes represent another of the various kinds of disbursements encountered in an economic analysis. The starting point in an after-tax computation is the before-tax cash flow. Generally, the before-tax cash flow contains three types of entries:

1. Disbursements of money to purchase capital assets. These expenditures create no direct tax consequence for they are the exchange of one asset (cash) for another (capital equipment).

2. Periodic receipts and/or disbursements representing operating income and/or expenses. These increase or decrease the year-by-year tax liability of the firm.

3. Receipts of money from the sale of capital assets, usually in the form of a salvage value when the equipment is removed. The tax consequence depends on the relationship between the book value (cost − depreciation taken) of the asset and its salvage value.

Situation	*Tax Consequence*
Salvage value > Book value	Capital gain on difference
Salvage value = Book value	No tax consequence
Salvage value < Book value	Capital loss on difference

After the before-tax cash flow, the next step is to compute the depreciation schedule for any capital assets. Next, taxable income is the taxable component of the before-tax cash flow minus the depreciation. Then, the income tax is the taxable income times the appropriate tax rate. Finally, the after-tax cash flow is the before-tax cash flow adjusted for income taxes.

To organize these data, it is customary to arrange them in the form of a cash flow table, as follows:

Year	*Before-tax cash flow*	*Depreciation*	*Taxable income*	*Income taxes*	*After-tax cash flow*
0	·		·	·	·
1	·	·	·	·	·
·	·	·	·	·	·

EXAMPLE 26

A corporation expects to receive \$32,000 each year for 15 years from the sale of a product. There will be an initial investment of \$150,000. Manufacturing and sales expenses will be \$8067 per year. Assume straight line depreciation, a 15-year useful life and no salvage value. Use a 46% income tax rate.

Determine the projected after-tax rate of return.

Solution

$$\text{Straight line depreciation} = \frac{P - S}{n} = \frac{150{,}000 - 0}{15}$$
$$= \$10{,}000 \text{ per year}$$

Year	*Before-tax cash flow*	*Depreciation*	*Taxable income*	*Income taxes*	*After-tax cash flow*
0	−150,000				−150,000
1	+23,933	10,000	13,933	−6,409	+17,524
2	+23,933	10,000	13,933	−6,409	+17,524
·	·	·	·	·	·
·	·	·	·	·	·
·	·	·	·	·	·
15	+23,933	10,000	13,933	−6,409	+17,524

Take the after-tax cash flow and compute the rate of return at which PW of cost equals PW of benefits.

$$150{,}000 = 17{,}524(P/A,i,15)$$
$$(P/A,i,15) = \frac{150{,}000}{17{,}524} = 8.559$$

From Compound Interest Tables, $i = 8\%$.

A CONCLUDING COMMENT

As you have seen, engineering economics is not a complex subject. There are, however, a group of fundamental concepts and techniques that must be mastered. This brief discussion has sought to provide a sound overview to these fundamentals. Unfortunately, there are likely to be other difficulties that you will encounter. In those situations, you will need to refer to one of the standard textbooks on the subject. I have listed my favorite reference (which should be no surprise) below.

Reference

Newnan, D. G. *Engineering Economic Analysis*, 3rd ed., 1988; Engineering Press, Inc., P.O. Box 1, San Jose, CA 95103-0001.

ENGR ECON 1

About how long will it take for $10,000 invested at 5% per year, compounded annually, to double in value?

(a) 5 yrs
(b) 10 yrs
(c) 15 yrs
(d) 20 yrs
(e) 25 yrs

SOLUTION

P = $10,000   F = $20,000   i = 0.05   n = unknown

Using the single payment compound amount factor

$F = P(1+i)^n \qquad 1.05^n = \frac{20{,}000}{10{,}000} = 2 \qquad n = 14.2 \text{ yrs}$ ●

Alternate solution using compound interest tables

F = P(F/P,5%,n)   (F/P,5%,n) = 20,000/10,000 = 2

From 5% interest tables: (F/P,5%,14) = 1.98   (F/P,5%,15) = 2.08

n = 14.2 ●

Answer is (c)

ENGR ECON 2

If $200 is deposited in a savings account at the beginning of each of 15 years and the account draws interest at 7% per year, compounded annually, the value of the account at the end of 15 years will be most nearly

(a) $5000
(b)   5400
(c)   6000
(d)   6900
(e)   7200

SOLUTION

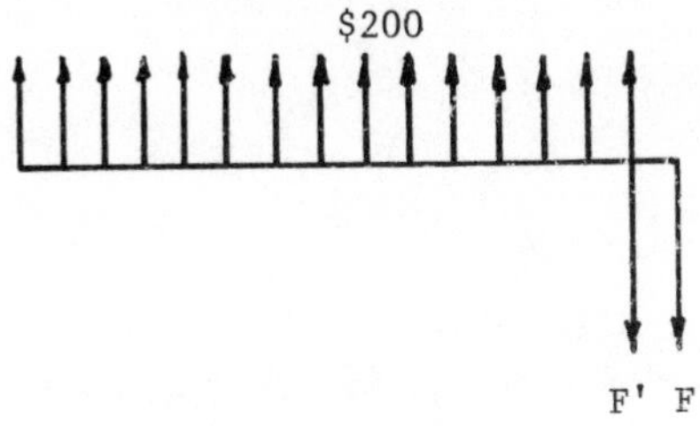

$F' = A(F/A,i\%,n) = \$200(F/A,7\%,15) = 200(25.129) = \$5025.80$

$F = F'(F/P,i\%,n) = 5025.80(F/P,7\%,1) = 5025.80(1.07) = \$5377.61$ ●

Answer is (b)

ENGR ECON 3

How many months at an interest rate of 1 percent per month does money have to be invested before it will double in value?

(a) 59 months
(b) 62
(c) 70
(d) 76
(e) 83

SOLUTION

Let $P = \$1$ $F = \$2$ $i = 0.01$ per month $n = $ number of months

$F = P(1 + i)^n$ $\$2 = \$1(1.01)^n$ $1.01^n = 2$

The solution may be obtained from a 1% compound interest table or by hand calculator. $n = 70$ months. ●

Answer is (c)

ENGR ECON 4

A department store charges one and one-half percent interest per month on credit purchases. This is equivalent to a nominal annual interest rate of

(a)   1.5 percent
(b)   15.0
(c)   18.0
(d)   19.6
(e)   21.0

SOLUTION

The nominal interest rate is the annual interest rate ignoring the effect of any compounding.

Nominal interest rate = $1\frac{1}{2}$ percent/month x 12 months = 18% ●

Answer is (c)

ENGR ECON 5

A bank charges $1\frac{1}{2}\%$ per month on the unpaid balance for purchases made with its credit card. This is equivalent to what effective annual interest rate?

(a) 1.5%
(b) 12%
(c) 18%
(d) 19.5%
(e) 39%

SOLUTION

Effective interest rate is the annual interest rate, taking into account the effect of any compounding during the year.

Effective interest rate = $(1 + i)^m - 1$ where i = interest rate/interest period.
m = number of compoundings per year.

$= (1 + 0.015)^{12} - 1$

$= 0.1956 = 19.56\%$ ●

Answer is (d)

ENGR ECON 6

A bank pays one percent interest on savings accounts four times a year. The effective annual interest rate is

(a) 1.00%
(b) 1.04%
(c) 3.96%
(d) 4.00%
(e) 4.06%

SOLUTION

Effective interest rate = $(1 + 0.01)^4 - 1 = 0.0406 = 4.06\%$ ●

Answer is (e)

ENGR ECON 7

What interest rate, compounded quarterly, is equivalent to a 9.31% effective interest rate?

(a) 2.25%
(b) 2.33%
(c) 4.66%
(d) 9.00%
(e) 9.31%

SOLUTION

$0.0931 = (1 + i)^4 - 1$

$1.0931^{0.25} = 1 + i$

$1.0225 = 1 + i$

i = 2.25% per quarter
i = 9% annual interest ●

Answer is (d)

ENGR ECON 8

A principal sum P is invested at a nominal interest rate r, compounded m times a year, for n years. The accumulated amount at the end of this period will be

(a) $P(1 + r/m)^{r}$

(b) $P(1 + r/m)^{m}$

(c) $P(1 + r/m)^{n/m}$

(d) $P(1 + r/m)^{m/r}$

(e) None of these

The relationship between a future sum F and a present sum P is

$F = P(1 + i)^{n}$ where i = interest rate per interest period
n = number of interest periods

The nominal annual interest rate r is the sum of the individual interest payments in a one year period. Therefore, the interest rate per interest period $i = r/m$. The number of interest periods in a particular situation is the number of interest periods per year m times the number of years n. Note that the problem specifies n years, while conventional notation defines n as the number of interest periods. Thus the problem makes the number of interest periods equal to the number of compoundings per year m times the number of years.

$i = r/m$ and no. of interest periods = mn

Substituting these into

$F = P(1 + i)^{n}$ gives $F = P(1 + r/m)^{mn}$ ●

Answer is (e)

ENGR ECON 9

In the formula $P = F(1 + i)^{-n}$ the factor $(1 + i)^{-n}$ is called the

(a) sinking fund factor
(b) single payment present worth factor
(c) single payment compound amount factor
(d) capital recovery factor
(e) uniform series present worth factor

Answer is (b) ●

ENGR ECON 10

A fund established to produce a desired amount at the end of a given period by means of a series of payments throughout the period is called a sinking fund, and is represented by the formula:

(a) $A = F\left[\frac{i}{(1 + i)^{n} - 1}\right]$

(b) $A = F\left[\frac{(1 + i)^{n}}{i}\right]$

(c) $A = P(1 + i)^{-n}$

(d) $A = P\left[\frac{(1 + i)^{n} - 1}{i(1 + i)^{n}}\right]$

(e) $A = F\left[\frac{i}{(1 + i)^{n}}\right]$

$$A = F\left[\frac{i}{(1 + i)^n - 1}\right] \bullet$$

Answer is (a)

ENGR ECON 11

Which of the following relationships between compound interest factors is NOT correct?

(a) Single payment compound amount factor and single payment present worth factor are reciprocals.

(b) Sinking fund factor and uniform series compound amount factor are reciprocals.

(c) Capital recovery factor and uniform series present worth factor are reciprocals.

(d) Capital recovery factor equals sinking fund factor plus the interest **rate**.

(e) Capital recovery factor and sinking fund factor are reciprocals.

Since answer (c) says Capital recovery = 1/(Series present worth)
and (e) says Capital recovery = 1/(Sinking fund)
it would necessarily follow that Series present worth = Sinking fund, and that obviously is not true. Thus the incorrect statement is either (c) or (e).

(a) $(F/P,i\%,n) = \frac{1}{(P/F,i\%,n)}$

(b) $(A/F,i\%,n) = \frac{1}{(F/A,i\%,n)}$

(c) $(A/P,i\%,n) = \frac{1}{(P/A,i\%,n)}$

(d) $(A/P,i\%,n) = (A/F,i\%,n) + i$

(e) $(A/P,i\%,n) \neq \frac{1}{(A/F,i\%,n)}$ ●

Answer is (e)

ENGR ECON 12

For some interest rate i and some number of interest periods n, the uniform series capital recovery factor is 0.0854 and the sinking fund factor is 0.0404. The interest rate i must be

(a) $3\frac{1}{2}\%$

(b) $4\frac{1}{2}\%$

(c) 6%

(d) 8%

(e) 9%

In Part (d) of ENGR ECON 11 we saw that

$$(A/P,i\%,n) = (A/F,i\%,n) + i$$

If we substitute the values given here into the equation, we have

$$0.0854 = 0.0404 + i$$

$$i = 0.0854 - 0.0404 = 0.045 = 4\tfrac{1}{2}\%$$ ●

Answer is (b)

ENGR ECON 13

An "annuity" is defined as

(a) Earned interest due at the end of each interest period.

(b) Cost of producing a product or rendering a service.

(c) Total annual overhead assigned to a unit of production.

(d) Amount of interest earned by a unit of principal in a unit of time.

(e) A series of equal payments occurring at equal periods of time.

Answer is (e) ●

ENGR ECON 14

An individual wishes to deposit a certain quantity of money now so that at the end of five years he will have \$500. With interest at 4% per year, compounded semiannually, how much must he deposit now?

(a) \$340.30
(b) 400.00
(c) 410.15
(d) 416.95
(e) 608.35

$P = F(1 + i)^{-n}$ where $F = \$500$.
$i = 0.02$ per interest period (i is *not* 0.04)
$n = 10$ interest periods

Calculator solution:

$$P = \$500(1 + 0.02)^{-10} = \$410.17$$ ●

Compound interest table solution:

$$P = \$500(P/F,2\%,10) = \$500(0.8203) = \$410.15$$ ●

Answer is (c)

ENGR ECON 15

The present worth of an obligation of \$10,000 due in 10 years if money is worth 9% is nearest to

(a) \$10,000
(b) 9,000
(c) 7,500
(d) 6,000
(e) 4,500

$P = F(P/F,9\%,10) = \$10,000(0.4224) = \$4224$ ●

Answer is (e)

ENGR ECON 16

$1000 is borrowed for one year at an interest rate of 1% per month. If this same sum of money is borrowed for the same period at an interest rate of 12% per annum, the saving in interest charges would be:

(a) $ 0
(b)   3
(c)   5
(d)   7
(e)   14

Calculator solution:

At $i = 1\%$ per month $F = \$1000(1 + 0.01)^{12} = \$1126.83$
At $i = 12\%$ per year $F = 1000(1 + 0.12)^{1} = \underline{1120.00}$

Saving in interest = $6.83 ●

Interest table solution:

At $i = 1\%$ per month $F = \$1000(F/P,1\%,12) = 1000(1.127) = \$1127.00$
At $i = 12\%$ per year $F = 1000(F/P,12\%,1) = 1000(1.120) = \underline{1120.00}$

Saving in interest = $7.00 ●

Answer is (d)

ENGR ECON 17

$$A = \$10,000$$

n = 15
i = 5%

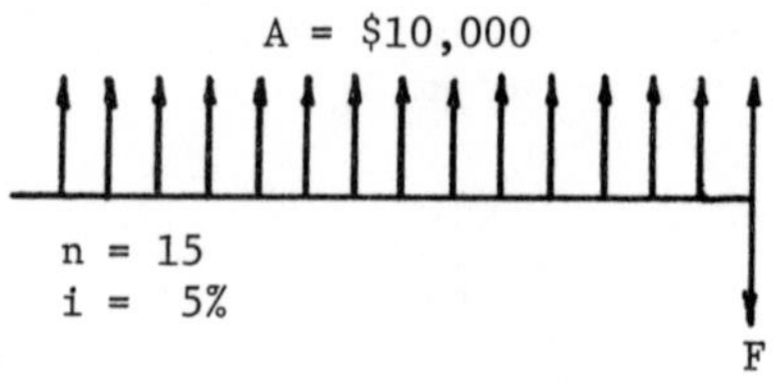

F is closest to

(a) $105,000
(b)   150,000
(c)   157,000
(d)   215,000
(e)   262,000

$F = \$10,000(F/A,5\%,15) = 10,000(21.579) = \$215,790$ ●

Answer is (d)

ENGR ECON 18

A A A A A A A A A A A

n = 12
i = 6%   F = $10,000

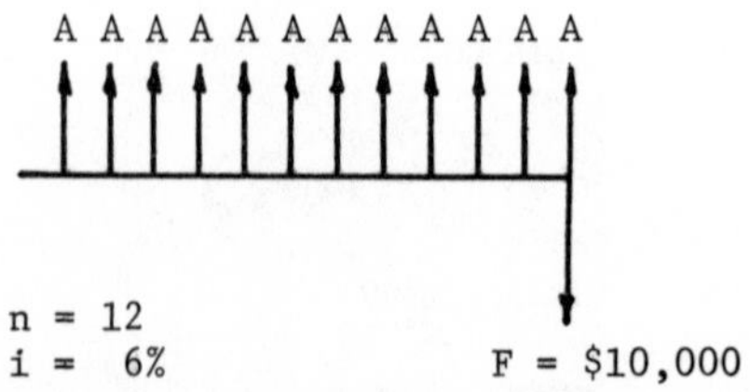

A is closest to

(a) $ 593
(b) 783
(c) 833
(d) 883
(e) 1193

A = 10,000(A/F,6%,12) = 10,000(0.0593) = $593 ●

Answer is (a)

ENGR ECON 19

A company deposits $1000 every year for ten years in a bank. The company makes no deposits during the subsequent five years. If the bank pays 8% interest, the amount in the account at the end of 15 years is nearest to

(a) $10,800
(b) 15,000
(c) 16,200
(d) 21,200
(e) 25,200

F = 1000(F/A,8%,10)(F/P,8%,5) = 1000(14.487)(1.469) = $21,281.40 ●

Answer is (d)

ENGR ECON 20

$25,000 is deposited in a savings account that pays 5% interest, compounded semiannually. Equal annual withdrawals are to be made from the account, beginning one year from now and continuing forever. The maximum amount of the equal annual withdrawal is closest to:

(a) $ 625
(b) 1000
(c) 1250
(d) 1265
(e) 1365

Effective interest rate = $(1 + 0.025)^2 - 1$ = 0.050625 = 5.0625%

Annual withdrawal A = Pi = 25,000(0.050625) = $1265.63 ●

Answer is (d)

ENGR ECON 21

What present sum would need to be put in a savings account now to provide a $1000 annual withdrawal for the next 50 years, if interest is 6%? The present sum is closest to:

(a) $ 1,000
(b) 10,000
(c) 25,000
(d) 37,500
(e) 50,000

P = $1000(P/A,6%,50) = 1000(15.762) = $15,762 ●

Answer is (b)

ENGR ECON 22

What present sum is equivalent to a series of $1000 annual end-of-year payments, if a total of 10 payments are made and interest is 6%?
The present sum is closest to

(a) $ 6,250
(b) 7,350
(c) 9,400
(d) 10,000
(e) 10,600

P = 1000(P/A,6%,10) = 1000(7.360) = $7360 ●

Answer is (b)

ENGR ECON 23

A woman made ten annual end-of-year purchases of $1000 of common stock. At the end of the tenth year she sold all the stock for $12,000.
What interest rate did she obtain on her investment?

(a) 2%
(b) 4%
(c) 8%
(d) 14%
(e) 20%

F = A(F/A,i%,n) 12,000 = 1000(F/A,i%,10)

$$(F/A,i\%,10) = \frac{12{,}000}{1{,}000} = 12$$

In the 4% interest table
(F/A,4%,10) = 12.006 i is very close to 4% ●

Answer is (b)

ENGR ECON 24

What present sum would be needed to provide for annual end-of-year payments of $15 each, forever? Assume interest is 8%

(a) $120.00
(b) 137.50
(c) 150.00
(d) 187.50
(e) 375.00

In the special situation where n equals infinity, the value of the capital recovery factor is i.

$$A = P(A/P,i\%,\infty) = Pi$$

$$P = \frac{A}{i} = \frac{\$15}{0.08} = \$187.50$$ ●

Answer is (d)

ENGR ECON 25

What amount of money deposited 50 years ago at 8% interest would now provide a perpetual payment of $10,000 per year? The amount is nearest to:

(a) $ 3,000
(b) 8,000
(c) 50,000
(d) 82,000
(e) 125,000

The amount of money needed now to begin the perpetual payments:

$$P' = \frac{A}{i} = \frac{10,000}{0.08} = \$125,000$$

The amount of money that would need to have been deposited 50 years ago at 8% interest is:

$$P = \$125,000(P/F,8\%,50) = 125,000(0.0213) = \$2662.50$$

Answer is (a)

ENGR ECON 26

A dam was constructed for \$200,000. The annual maintenance cost is \$5000. If interest is 5%, the capitalized cost of the dam, including maintenance, is:

(a) \$100,000
(b) 200,000
(c) 215,000
(d) 250,000
(e) 300,000

Capitalized cost is defined as the present worth of perpetual service, and is frequently used in connection with public works projects.

$$\text{Capitalized cost} = \$200,000 + \frac{A}{i} = \$200,000 + \frac{\$5000}{0.05} = \$300,000$$

Answer is (e)

ENGR ECON 27

Given a sum of money *Q* that will be received six years from now. At 5% annual interest the present worth now of *Q* is \$60. At this same interest rate, what would be the value of *Q* ten years from now?

(a) \$ 60.00
(b) 76.58
(c) 90.00
(d) 97.74
(e) 120.00

This problem illustrates the concept of equivalence. The present sum P = \$60 is equivalent to *Q* six years hence at 5% annual interest.
The future sum F may be calculated by either of two methods:

$$F = Q(F/P,5\%,4) \quad \text{Eqn (1)}$$

or $$F = P(F/P,5\%,10) \quad \text{Eqn (2)}$$

Since P is known, Eqn 2 may be solved directly.

$$F = P(F/P,5\%,10) = \$60(1.629) = \$97.74$$

Answer is (d)

ENGR ECON 28

A piece of property is purchased for \$10,000 and yields a \$1000 yearly profit. If the property is sold after five years, what is the minimum price to break-even, with interest at 6%?

(a) \$ 5,000
(b) 6,554
(c) 7,743
(d) 8,314
(e) 10,000

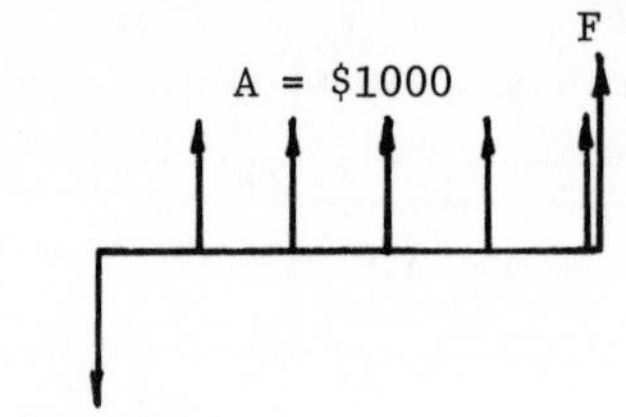

F = $10,000(F/P,6%,5) - $1000(F/A,6%,5)
= $10,000(1.338) - $1000(5.637) = 13,380 - 5637 = $7743 ●

Answer is (c)

ENGR ECON 29

A steam boiler is purchased on the basis of guaranteed performance. A test indicates that the operating cost will be $300 more per year than the manufacturer guaranteed. If the expected life of the boiler is 20 years and money is worth 8%, how much should the purchaser deduct from the purchase price to compensate for the extra operating cost?

(a) $2945
(b) 3320
(c) 4102
(d) 5520
(e) 6000

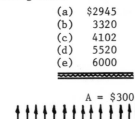

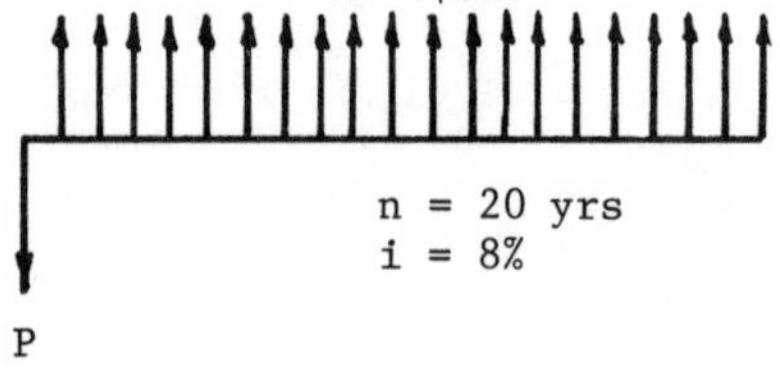

P = $300(P/A,8%,20) = 300(9.818) = $2945.40 ●

Answer is (a)

ENGR ECON 30

Annual maintenance costs for a particular section of highway pavement are $2000. The placement of a new surface would reduce the annual maintenance cost to $500 per year for the first five years and to $1000 per year for the next five years. The annual maintenance after ten years would again be $2000. If maintenance costs are the only saving, what maximum investment can be justified for the new surface? Assume interest at 4%.

(a) $ 5,500
(b) 7,170
(c) 10,000
(d) 10,340
(e) 12,500

Benefits = $1500 per year for the first five years and
$1000 per year for the subsequent five years.

There are several ways of computing the present worth of benefits. Two solutions will be presented.

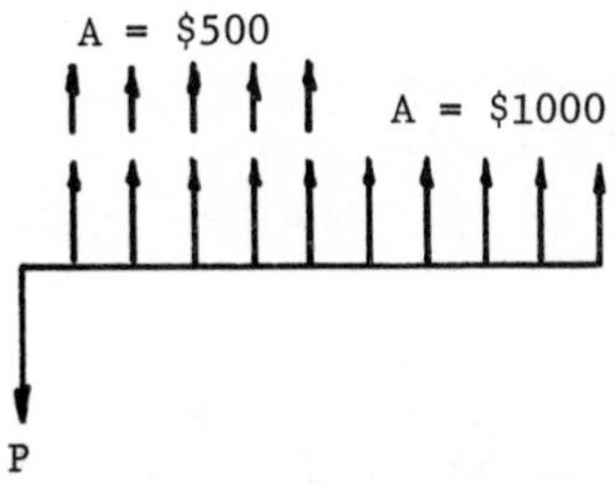

As the sketch indicates, the benefits may be considered as $1000 per year for ten years, plus an additional $500 benefit in each of the first five years.

Maximum investment = Present Worth of benefits
= $1000(P/A,4%,10) + $500(P/A,4%,5)
= $1000(8.111) + $500(4.452)
= $10,337 ●

Alternate Solution.

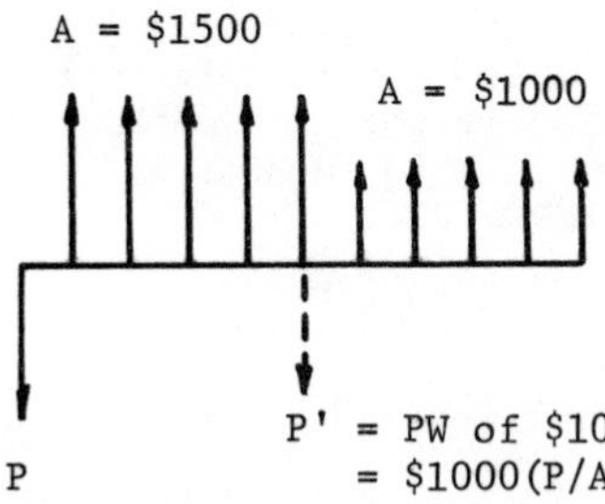

P' = PW of $1000 per year for five years
= $1000(P/A,4%,5) = 1000(4.452) = $4452

Maximum investment = Present Worth of Benefits
= $1500(P/A,4%,5) + P'(P/F,4%,5)
= $1500(4.452) + $4452(0.8219)
= $10,337 ●

Answer is (d)

ENGR ECON 31

A man buys a small garden tractor. There will be no maintenance cost the first year as the tractor is sold with one year's free maintenance. The second year the maintenance is estimated at $20. In subsequent years the maintenance cost will increase $20 per year (that is, 3rd year maintenance will be $40; 4th year maintenance will be $60, and so forth). How much would need to be set aside now at 5% interest to pay the maintenance costs on the tractor for the first six years of ownership?

(a) $101.52
(b) 164.74
(c) 239.36
(d) 284.13
(e) 300.00

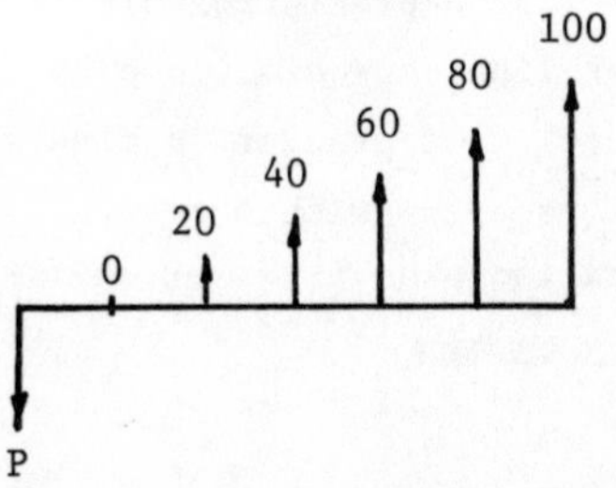

Solution using single payment present worth factors

$P = 20(P/F,5\%,2) + 40(P/F,5\%,3) + 60(P/F,5\%,4) + 80(P/F,5\%,5) + 100(P/F,5\%,6)$

$\quad = 20(0.9070) + 40(0.8638) + 60(0.8227) + 80(0.7835) + 100(0.7462)$

$\quad = \$239.35$ ●

Alternate solution using the gradient present worth factor

$P = 20(P/G,5\%,6) = 20(11.968) = \$239.36$ ●

Answer is (c)

ENGR ECON 32

Which one of the following is *NOT* a method of depreciating plant equipment for accounting and engineering economic analysis purposes?

    (a)   Double entry method
    (b)   Fixed percentage method
    (c)   Sum-of-years digits method
    (d)   Straight line method
    (e)   Sinking fund method

Answer is (a) ●

ENGR ECON 33

A manufacturing company buys a machine for \$50,000. It estimates the machine's useful life is 20 years and that it can then be sold for \$5000. Using straight line depreciation, what is the annual depreciation charge?

    (a)   \$2000
    (b)   2250
    (c)   2500
    (d)   2750
    (e)   3000

Straight line depreciation

$$\text{Annual depreciation charge} = \frac{P - S}{n} = \frac{50,000 - 5,000}{20} = \$2250 \ \bullet$$

Answer is (b)

ENGR ECON 34

In determining the average annual cost of a proposed project, the formula

$$\frac{P - S}{n} + (P - S)\left(\frac{i}{2}\right)\left(\frac{n + 1}{n}\right) + Si \quad \text{represents the economic method of}$$

    (a)   Sinking fund depreciation plus interest on first cost.

    (b)   Straight line depreciation plus interest on first cost.

    (c)   Straight line depreciation plus average interest.

    (d)   Capital recovery with a return.

    (e)   Amortization plus interest on first cost.

Straight line depreciation $= \frac{P - S}{n}$

Average interest $= \frac{\text{First year's interest + Last year's interest}}{2}$

First year's interest = (P - S)i + Si

Last year's interest $= \frac{(P - S)}{n}(i) + Si$

$$\text{Average interest} = \frac{(P - S)i + Si + \frac{(P - S)}{n}(i) + Si}{2}$$

$$= (P - S)\left(\frac{i}{2}\right)\left(\frac{n + 1}{n}\right) + Si$$

Therefore, straight line depreciation plus average interest equals

$$\frac{P - S}{n} + (P - S)\left(\frac{i}{2}\right)\left(\frac{n + 1}{n}\right) + Si$$ ●

Answer is (c)

ENGR ECON 35

Company A has fixed expenses of \$15,000 per year and each unit of product has a \$0.002 variable cost. Company B has fixed expenses of \$5000 per year and can produce the same unit of product at a \$0.05 variable cost. At what number of units of annual production will Company A have the same overall cost as Company B? Quantity is nearest to

(a) 100,000 units
(b) 200,000
(c) 300,000
(d) 400,000
(e) 2,000,000

Let x = annual production (units)

Total cost to Company A = Total cost to Company B

$$15{,}000 + 0.002x = 5000 + 0.05x$$

$$x = \frac{10{,}000}{0.048} = 208{,}333 \text{ units}$$ ●

Answer is (b)

ENGR ECON 36

Plan *A* requires a \$100,000 investment now.
Plan *B* requires an \$80,000 investment now and an additional \$40,000 investment later. At 8% interest, what is the breakeven point on the timing of the additional \$40,000 later?

(a) 3 years
(b) 5
(c) 7
(d) 9
(e) 11

The difference between the alternatives is that Plan *A* requires \$20,000 extra now and Plan *B* requires \$40,000 extra n years hence.

At breakeven

$$\$20,000 = \$40,000(P/F,8\%,n)$$

$$(P/F,8\%,n) = 0.5$$

From the 8% compound interest table (P/F,8%,9) = 0.5002
Therefore, n = 9 years. ●

Answer is (d)

ENGR ECON 37 *(This is a 10-question problem)*

A company is considering buying a new piece of machinery. Two models are available.

	Machine I	*Machine II*
Initial cost	\$80,000	\$100,000
End of useful life salvage value, S	20,000	25,000
Annual operating cost	18,000	15,000 first 10 years; 20,000 thereafter
Useful life	20 yrs	25 yrs

Answer the following ten questions, based on a 10% interest rate.

QUESTION 1 What is the equivalent uniform annual cost for Machine I? The annual cost is closest to

(a) \$21,000
(b) 23,000
(c) 25,000
(d) 27,000
(e) 29,000

$$EUAC = (P - S)(A/P,i\%,n) + Si + \text{Annual operating cost}$$
$$= (80,000 - 20,000)(A/P,10\%,20) + 20,000(0.10) + 18,000$$
$$= 60,000(0.1175) + 2000 + 18,000$$
$$= 27,050$$ ●

Answer is (d)

QUESTION 2 What is the equivalent uniform annual cost for Machine II? The annual cost is closest to

(a) \$21,000
(b) 23,000
(c) 25,000
(d) 27,000
(e) 29,000

$$EUAC = (100,000 - 25,000)(A/P,10\%,25) + 25,000(0.10)$$
$$+ 20,000 - 5000(P/A,10\%,10)(A/P,10\%,25)$$
$$= 75,000(0.1102) + 2500 + 20,000 - 5000(6.145)(0.1102)$$
$$= \$27,379$$ ●

Answer is (d)

QUESTION 3 The capitalized cost based on Machine I is closest to

(a) $ 60,000
(b) 80,000
(c) 230,000
(d) 270,000
(e) 420,000

Capitalized cost = Present worth of an infinite life.
In Question 1 we computed the Equivalent Uniform Annual Cost (EUAC).

$$P = \frac{A}{i}$$

$$\text{Capitalized Cost} = \frac{A}{i} = \frac{\text{EUAC}}{i} = \frac{\$27,050}{0.10} = \$270,500$$ ●

Answer is (d)

QUESTION 4 If Machine I is purchased and a fund is set up to replace Machine I at the end of 20 years, the uniform annual deposit that should be made to the fund is nearest to:

(a) $1000
(b) 2000
(c) 3000
(d) 4000
(e) 5000

Required future sum F = $80,000 - 20,000 = $60,000

Annual deposit A = $60,000(A/F,10%,20) = $60,000(0.0175) = $1050 ●

Answer is (a)

QUESTION 5 Machine I will produce an annual savings in material of $25,700 a year. What is the before-tax rate of return if Machine I is installed? The rate of return is closest to:

(a) 6%
(b) 8%
(c) 10%
(d) 20%
(e) 35%

The cash flow for this situation is:

Year	Cash Flow
0	-$80,000
1-20	+25,700 -18,000
20	+20,000

Write one equation with i as the only unknown.

80,000 = (25,700 - 18,000)(P/A,i%,20) + 20,000(P/F,i%,20)

Try i = 8%

80,000 = 7700(9.818) + 20,000(0.2145) = 79,888

Therefore, the rate of return is very close to 8% ●

Answer is (b)

QUESTION 6 Assuming Sum-Of-Years Digits depreciation, what would be the book value of Machine I after two years? Book value is closest to:

(a) $21,000
(b) 42,000
(c) 59,000
(d) 69,000
(e) 79,000

Sum-Of-Years Digits depreciation:

$$\text{Depreciation charge in any year} = \frac{\text{Remaining useful life at beginning of year}}{\text{Sum-Of-Years Digits for total useful life}}(P - S)$$

$$\text{Sum-Of-Years Digits} = \frac{n}{2}(n + 1) = \frac{20}{2}(21) = 210$$

$$\text{1st year depreciation} = \frac{20}{210}(80{,}000 - 20{,}000) = \$\ 5{,}714$$

$$\text{2nd year depreciation} = \frac{19}{210}(80{,}000 - 20{,}000) = \underline{5{,}429}$$

$$\text{TOTAL} = \$11{,}143$$

Book Value = Cost - Depreciation to date
= $80,000 - 11,143
= $68,857

Answer is (d)

QUESTION 7 Assuming Double Declining Balance depreciation, what would be the book value of Machine II after 3 years?

(a) $16,000
(b) 22,000
(c) 58,000
(d) 78,000
(e) 83,000

Double Declining Balance depreciation:

$$\text{Depreciation charge in any year} = \frac{2}{n}(P - \text{Depreciation charges to date})$$

$$\text{1st year depreciation} = \frac{2}{25}(100{,}000 - 0) = \$8{,}000$$

$$\text{2nd year depreciation} = \frac{2}{25}(100{,}000 - 8{,}000) = 7{,}360$$

$$\text{3rd year depreciation} = \frac{2}{25}(100{,}000 - 15{,}360) = \underline{6{,}771}$$

$$\text{TOTAL: } \$22{,}131$$

Book Value = Cost - Depreciation to date
= $10,000 - 22,131
= $77,869

Answer is (d)

QUESTION 8 A new building is being considered to house some equipment. The new building will reduce maintenance costs by $6000 per year for the first ten years, and $3000 per year thereafter. Based on a 50-year analysis period, what building construction cost can be justified? The justified construction cost is closest to:

(a) $ 30,000
(b)   50,000
(c)   90,000
(d)   140,000
(e)   180,000

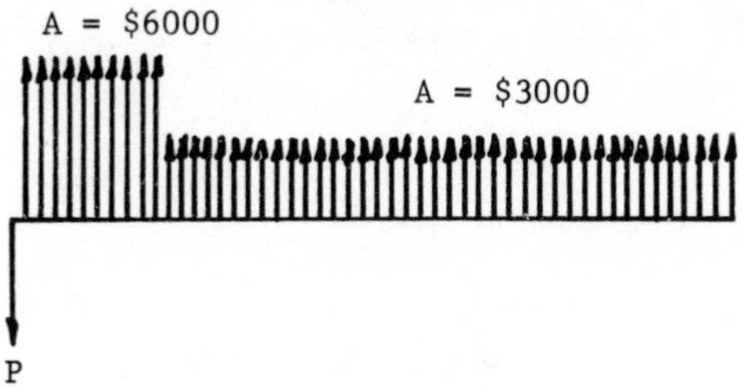

The justified construction cost = Present Worth of savings

P = $6000(P/A,10%,10) + $3000(P/A,10%,40)(P/F,10%,10)

= $6000(6.145) + $3000(9.779)(0.3855)

= $48,179 ●

Alternate solution:

P = $3000(P/A,10%,50) + $3000(P/A,10%,10)

= $3000(9.915) + $3000(6.145)

= $48,180 ●

Answer is (b)

QUESTION 9 The manufacturers of Machine II have announced a price reduction for the machine. What is the breakeven initial price for Machine II, when compared to Machine I?

(a) $ 91,000
(b)   94,000
(c)   97,000
(d)   100,000
(e)   103,000

In *Question 1* the equivalent uniform annual cost for Machine I was computed to be $27,050.
For breakeven set the EUAC of Machine I equal to the EUAC of Machine II, and compute initial cost of Machine II at this point.

$EUAC_I = EUAC_{II}$

$27,050 = (P - 25,000)(A/P,10%,25) + 25,000(0.10) + 20,000
$\qquad$ - 5000(P/A,10%,10)(A/P,10%,25)

$\qquad$ = (P - 25,000)(0.1102) + 22,500 - 5000(6.145)(0.1102)

$\qquad$ = 0.1102P - 2755 + 22,500 - 3386

$$\$27{,}050 = 0.1102P + 16{,}359$$

$$P = \frac{\$27{,}050 - 16{,}359}{0.1102} = \$97{,}015$$

Answer is (c)

QUESTION 10 What must be the salvage value of Machine I at the end of 20 years for the machine to have an equivalent uniform annual cost of $26,500?

(a) $10,000
(b) 20,000
(c) 30,000
(d) 40,000
(e) 50,000

We can write one equation with S as the only unknown.

$$\$26{,}500 = \$80{,}000(A/P,10\%,20) + 18{,}000 - S(A/F,10\%,20)$$
$$= \$80{,}000(0.1175) + 18{,}000 - S(0.0175)$$
$$= \$9400 + 18{,}000 - 0.0175S$$

$$S = \frac{\$27{,}400 - \$26{,}500}{0.0175} = \$51{,}429$$

Answer is (e)

ENGR ECON 38

A company uses 8000 wheels per year in its manufacture of golf carts. The wheels cost $15 each and are purchased from an outside supplier. The money invested in the inventory costs 10% per year, and the warehousing cost amounts to an additional 2% per year. It costs $150 to process each purchase order.

The quantity of wheels that should be ordered each time an order is placed is nearest to

(a) 1200
(b) 1600
(c) 2000
(d) 2500
(e) 3000

SOLUTION

The simplest model for the Economic Order Quantity is

$$EOQ = \sqrt{\frac{2BD}{E}}$$

where
B = Ordering cost, $/order
D = Demand per period, units
E = Inventory holding cost, $/unit/period
EOQ = Economic order quantity, units

$$EOQ = \sqrt{\frac{2 \times \$150 \times 8000}{(10\% + 2\%)(15.00)}} = 1155 \text{ wheels}$$

Answer is (a)

ENGR ECON 39

A bank pays 10% nominal annual interest on special three year certificates. Answer the following three questions.

QUESTION 1 If the interest is compounded every three months, the effective annual interest rate is nearest to

(a) 10.000%
(b) 10.375
(c) 10.500
(d) 10.750
(e) 30.000

SOLUTION

$$\text{Effective } i = (1 + i)^m - 1 = (1 + 0.025)^4 - 1$$
$$= 0.1038 = 10.38\%$$

Answer is (b)

QUESTION 2 If interest is compounded daily, the effective annual interest rate is nearest to

(a) 10.000%
(b) 10.375
(c) 10.500
(d) 10.750
(e) 30.000

SOLUTION

$$\text{Effective } i = (1 + i)^m - 1 = \left[1 + \frac{0.10}{365}\right]^{365} -1$$
$$= 0.10516 = 10.516\%$$

Answer is (c)

QUESTION 3 If interest is compounded continuously, the effective annual interest rate is nearest to

(a) 10.000%
(b) 10.375
(c) 10.500
(d) 10.750
(e) 30.000

SOLUTION

For continuous compounding

Effective $i = e^r - 1$ where r = nominal interest rate

$= e^{0.10} - 1 = 0.10517 = 10.517\%$

Answer is (c)

ENGR ECON 40

Calculate the rate of return for the following investment opportunity.

Invest $100 now.
Receive two payments of $109.46 - one at the end of year 3 and one at the end of year 6.

The rate of return is nearest to

(a) 10%
(b) 20
(c) 30
(d) 40
(e) 50

SOLUTION

Set the Present Worth of cost = Present Worth of benefits

$$100 = 109.46(P/F,i\%,3) + 109.46(P/F,i\%,6)$$

Solve by trial and error:

Try i = 15%

$$100 \stackrel{?}{=} 109.46(0.6575) + 109.46(0.4323)$$

$$\stackrel{?}{=} 109.46(1.0898) = 119.29$$

The PW of benefits is greater than the PW of cost, indicating that the interest rate i is too low.

Try i = 20%

$$100 \stackrel{?}{=} 109.46(0.5787) + 109.46(0.3349)$$

$$\stackrel{?}{=} 109.46(0.9136) = 100.00$$

Here the PW of cost = PW of benefits when i = 20%.

Therefore, the rate of return = 20%

In most situations a trail i will not result in PW of Cost equal to the PW of benefits. A linear interpolation for i is usually required.

Answer is (b)

ENGR ECON 41

A machine part, operating in a corrosive atmosphere, is made of low-carbon steel, costs \$350 installed and lasts six years. If the part is treated for corrosion resistance it will cost \$700 installed.

How long must the treated part last to be as good an investment as the untreated part, if money is worth 7%?

(a) 8 years
(b) 11
(c) 14
(d) 17
(e) 20

SOLUTION

The Equivalent Uniform Annual Cost ("annual cost") of the untreated part is:

$$\$350(A/P,7\%,6) = 350(0.2098) = \$73.43$$

For breakeven, the annual cost of the treated part must be this same value, or

$$\$73.43 = \$700(A/P,7\%,n)$$

$$(A/P,7\%,n) = \frac{73.43}{700} = 0.1049$$

From compound interest tables:

n	(A/P,7%,n)
16 yrs	0.1059
17 yrs	0.1024

By linear interpolation we get

$$n = 16 + (1)\,\frac{0.1059 - 0.1049}{0.1059 - 0.1024} = 16 + \frac{0.0010}{0.0035} = 16.3 \text{ years}$$

Thus the treated part must last 16.3 years to be as good an investment as the untreated part, that is, breakeven is 16.3 years.

Answer is (d)

ENGR ECON 42

An investor is considering buying a 20-year corporate bond. The bond has a face value of \$1000 and pays 6% interest per year in two semiannual payments. Thus the purchaser of the bond would receive \$30 every 6 months, and in addition he would receive \$1000 at the end of 20 years, along with the last \$30 interest payment.

If the investor though he should receive 8% annual interest, compounded semiannually, the amount he would be willing to pay for the bond is closest to

(a) \$500
(b) \$600
(c) \$700
(d) \$800
(e) \$900

SOLUTION

The investor would be willing to pay the present worth of the future benefits, computed at 8% compounded semiannually.

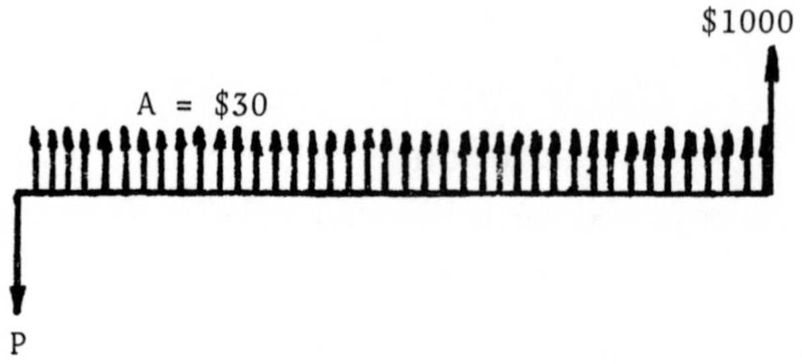

For semiannual interest periods

$i = 4\%$ per interest period

$n = 40$ interest periods

$$\text{Present Worth} = 30(P/A,4\%,40) + 1000(P/F,4\%,40)$$
$$= 30(19.793) + 1000(0.2083)$$
$$= 593.79 + 208.30$$
$$= \underline{\$802.09}$$

Answer is (d)

10

Typical Examination

This chapter gives a typical set of problems that can be considered fairly equivalent to one of the two four-hour sessions of the NCEE Electrical Engineering exam.

PROBLEM 1 **LOGIC & COMPUTERS**

Refer to the 10-part multiple choice problem in Chapter 8.

PROBLEM 2 **CONTROL SYSTEMS.** (Many of the NCEE questions ask something about system stability for a linear control system.)

The following system is to be stabilized by the addition of tachometer feedback, K_t.

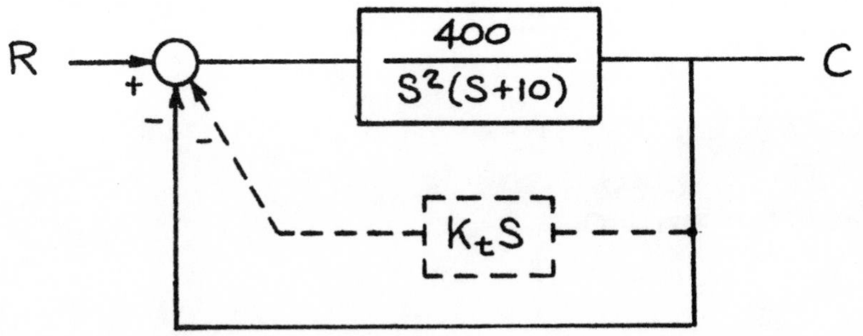

REQUIRED:

(a) Find the minimum value of K_t such that the system will just be stable.

(b) If the value of K_t (found in Part (a) were increased by a factor of 1.25 [i.e., 1.25 x K_t min], determine the

approximate step response characteristics.
(Hint: one method of an approximate solution is to use only a reasonably accurate root-locus sketch and then approximate with "standardized" 2nd order curves.)

SOLUTION

(a)

For stability use Routh-Hurwitz criterion for the characteristic polynominal:

$$G_{SYST} = \frac{C}{R} = \frac{400}{S^2(S+10) + (1+K_tS)400}$$

Since equivalent block diagram may be given as:

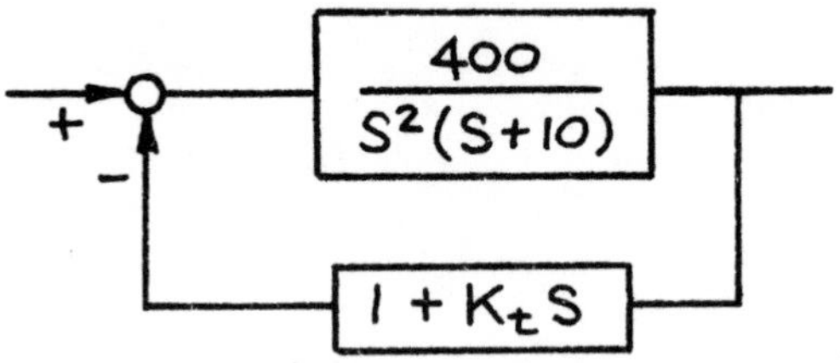

Characteristic Polynomial $= S^3 + 10S^2 + 400K_tS + 400$

Routh-Hurwitz array:

S^3	1	$400K_t$
S^2	10	400
S^1	X_1	

$$X_1 = \frac{(10)(400K_t) - (1)(400)}{10}$$

$$\therefore K_t > 0.1 \text{ for } X_1 > 0$$

$$(K_t \text{ minimum} = 0.1)$$

(b)

Plot root-locus for HG:

Let $H = 1 + K_tS = 1 + 0.125S = 0.125(8+S)$

$$\therefore HG = \frac{(0.125)(400)(S+8)}{S^2(S+10)} = \frac{50(S+8)}{S^2(S+10)}$$

Here K(root-locus) = 50.

Since solution is only approximate, let $\ell_1 \approx \ell_0$

and $\ell_2^2 = \omega_n^2$

Then zeta line $\approx 80^\circ$ and $\omega_n = \sqrt{50}$.

Zeta ≈ 0.17

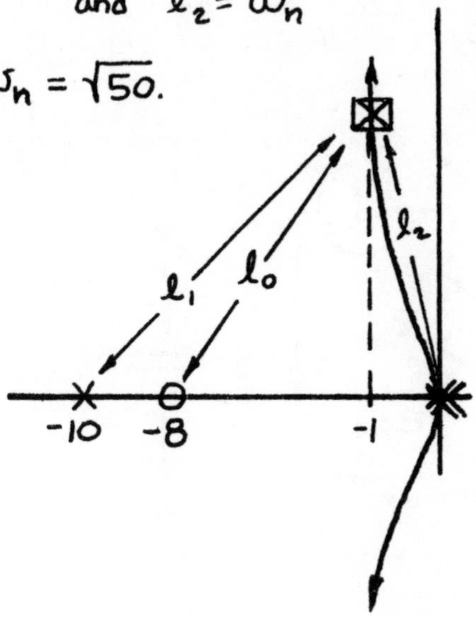

From any standard 2nd order transient response curves, one can easily determine percent overshoot and time to first peak, t_p.

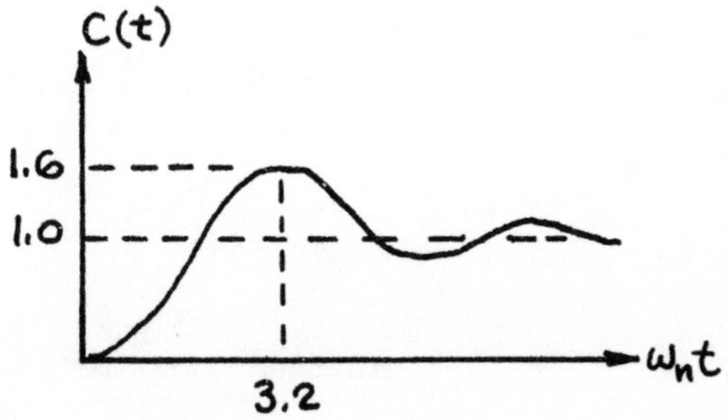

% O.S. = 60 %

$$t_p = \frac{3.2}{\omega_n} = 0.45 \text{ seconds}$$

PROBLEM 3 **ELECTRONICS.** (A typical question involves the small signal transistor or vacuum tube equivalent circuit amplifier - usually one stage.)

The following questions apply to the circuits 1 and 2 and their common graphical operation characteristic diagram shown on the following page.

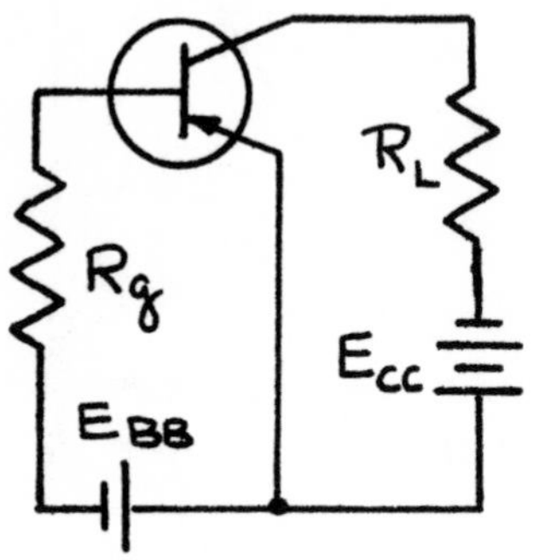

CKT 1

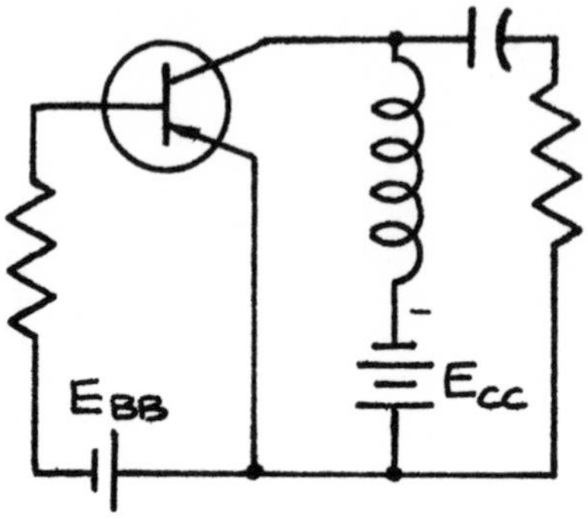

CKT 2

REQUIRED:

(a) What are the types and classes (mode) of amplifiers shown?

(b) What are their power outputs?

(c) What are their transistor-power dissipation loads during no-signal and maximum signal conditions?

(d) What are their collector efficiencies while under maximum conditions?

(e) What are their per cent of second harmonic distortion under maximum conditions?

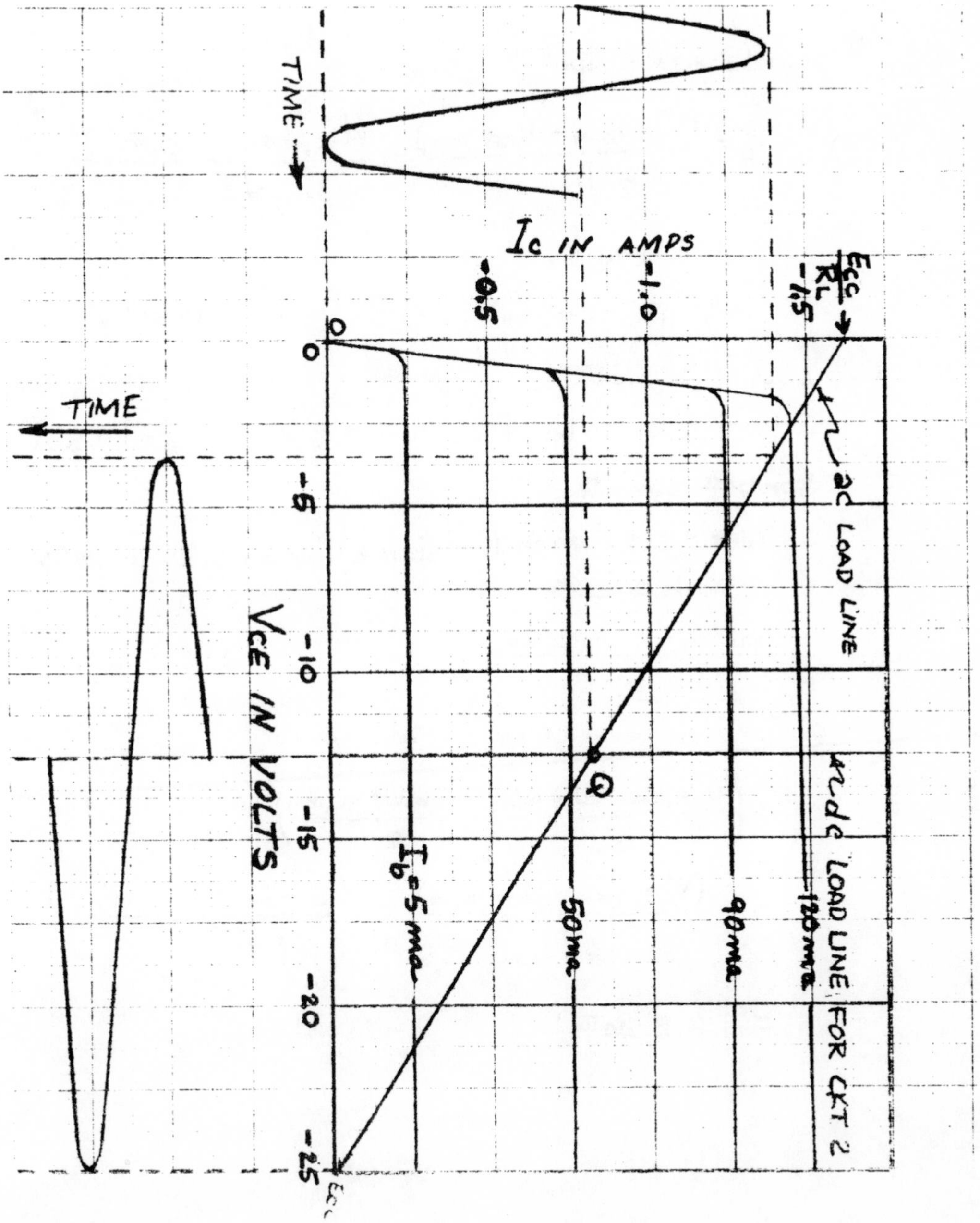
TIME
I_c IN AMPS
0
-0.5
-1.0
-1.5
$\frac{E_{CC}}{R_L}$
0
-5
-10
-15
-20
-25
V_{CE} IN VOLTS
TIME
ac LOAD LINE
dc LOAD LINE FOR CKT 2
Q
I_b=5 ma
50 ma
90 ma
120 ma

SOLUTION

(a) Common emitter, Class A amplifiers because the collector current flows during the entire cycle.

(b) Signal power output,

$$P_o = \frac{(v_{CE,max} - v_{CE,min})}{2\sqrt{2}} \times \frac{(i_{c,max} - i_{c,min})}{2\sqrt{2}}$$

$$= 3.76 \text{ watts}$$

The signal power output is the same for both circuits.

(c) No signal condition:

Q point given to be at $V_{CQ} = -12.5$ v, $I_{CQ} = -0.8$ A for both circuits.

∴ Power dissipation in transistor = $12.5 \times 0.8 = 10$ watts for both circuits.

Maximum signal condition:

(i) Circuit 1:

Average power dissipated in collector

$$= \left(\frac{v_{CE,max} + v_{CE,min}}{2}\right)\left(\frac{i_{c,max} + i_{c,min}}{2}\right) - \frac{(v_{CE,max} - v_{CE,min})}{2\sqrt{2}} \times \frac{(i_{c,max} - i_{c,min})}{2\sqrt{2}}$$

$$= 6.2 \text{ watts}$$

(ii) Circuit 2:

Average power dissipated in transistor

$$= \left(\frac{v_{CE,max} + v_{CE,min}}{2}\right)\left(\frac{i_{c,max} + i_{c,min}}{2}\right) - \frac{(v_{CE,max} - v_{CE,min})}{2\sqrt{2}} \times \frac{(i_{c,max} - i_{c,min})}{2\sqrt{2}}$$

$$= 9.975 \text{ watts} - 3.76 \text{ watts}$$

$$\simeq 6.2 \text{ watts}$$

(d) Collector efficiency

Circuit 1: P_o = ac power output to load = 3.76 watts

P_i = Power input = $V_{CC}\, I_{C,average}$

= 25 x 0.7 = 17.5 watts

$$\therefore \text{Collector Efficiency, } \eta_c = \frac{3.76}{17.5} \times 100 = 21.5\%$$

Circuit 2: AC power output is the same, P_o = 3.76 watts

Power input is reduced.

$$V_{CC} = \frac{V_{CE,max} + V_{CE,min}}{2} = 14.25 \text{ volts}$$

$$\therefore \text{Power input, } P_i = V_{CC}\, I_{c,average}$$

$$= 14.25 \times 0.7 = 9.975 \text{ watts}$$

$$\therefore \text{Collector Efficiency, } \eta_c = \frac{3.76}{9.975} \times 100 = 37.7\%$$

The collector efficiency is increased. This increase in efficiency is due to the fact that there is no dc power dissipated in the load.

(e)

Let $i_c(t) = I_o + I_1 \cos \omega t + I_2 \cos 2\omega t$

at $t=0$, $\quad i_c = i_{c,max} = I_o + I_1 + I_2 \quad (1)$

at $\omega t = \frac{\pi}{2}$, $\quad i_c = I_{CQ} = I_o - I_2 \quad (2)$

at $\omega t = \pi$, $\quad i_c = i_{c,min} = I_o - I_1 + I_2 \quad (3)$

From Eqns (1) and (3)

$$I_1 = \frac{i_{c,max} - i_{c,min}}{2} \quad (4)$$

From Eqns (2) and (3)

$$I_2 = \frac{i_{c,min} - I_{CQ} + I_1}{2}$$

$$= \frac{i_{c,min} - I_{CQ} + \frac{i_{c,max}}{2} - \frac{i_{c,min}}{2}}{2}$$

$$= \frac{i_{c,max} + i_{c,min} - 2I_{CQ}}{4} \quad (5)$$

∴ Percentage Second Harmonic Distortion

$$= \left|\frac{I_2}{I_1}\right| \times 100 = \frac{|-1.4 + 0 + 1.6|}{2\,|(-1.4 - 0)|} \times 100$$

$$= 7.1\%$$

PROBLEM 4 **POWER & SYSTEMS.** (At least one problem is usually included that involves power factor correction - either by capacitors or using a synchronous motor.)

Originally it is planned to furnish a plant load requirement of 1000 H.P. at 2200 volt, 3-phase, by induction motors operating at 80% power factor and 90% efficiency.

REQUIRED:

(a) Find the line current necessary to supply this load and generator capacity.

(b) Assume that rather than supplying the 1000 H.P. by induction motors, it is decided to produce 400 H.P. of this load by a synchronous motor operating in the over-excited leading mode of 85% (assume same efficiency as for an induction motor). Find the new total line current requirement and the overall power factor.

(c) If rather than installing the 400 H.P. synchronous motor (in Part b) it is considered feasible to use power factor correcting capacitors for the 1000 H.P. motors in Part (a) to achieve the same power factor correcting as obtained in Part (b). Determine the size (KVAR) of the capacitors needed.

SOLUTION

(a) Power input to motors:

$$\frac{1000}{0.9} = 1111 \text{ HP} \cong 830 \text{ KW}$$

Generator capacity

$$\frac{830 \text{ KW}}{0.8} = 1036 \text{ K.V.A.}$$

Current requirement

$$\frac{1036 \times 1000}{\sqrt{3} \times 2200} = 272 \text{ A}$$

(b) Power requirements:

Induct. motor input

$$\frac{600 \text{ HP}}{0.9} \times \frac{746}{1000} = 497 \text{ KW}$$

Induct. motor current

$$\frac{497 \times 1000}{\sqrt{3} \times 2200 \times 0.8} = 163 \text{ A.}$$

Synch. motor input

$$\frac{400 \text{ HP}}{0.9} \times \frac{746}{1000} = 332 \text{ kW}$$

Synch. motor current

$$\frac{332 \times 1000}{\sqrt{3} \times 2200 \times 0.85} = 102 \text{ A.}$$

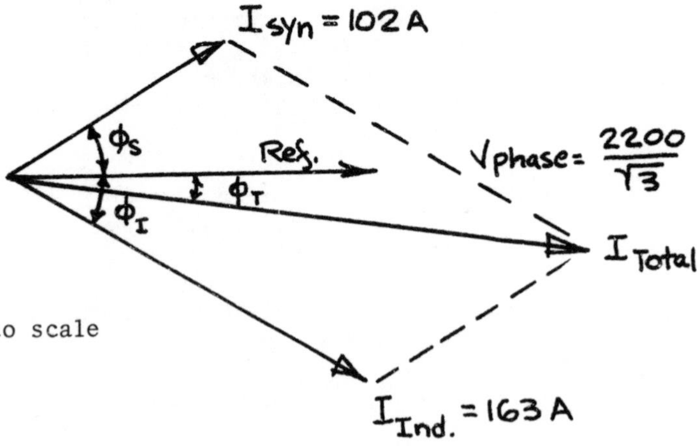

Not to scale

$$I_{Total} = \sqrt{(I_I \cos \phi_I + I_s \cos \phi_s)^2 + (I_I \sin \phi_I - I_s \sin \phi_s)^2}$$

$$= \sqrt{(163 \times 0.8 + 102 \times 0.85)^2 + (163 \times 0.6 - 102 \times 0.53)^2}$$

$$= 221 \text{ A } \angle \phi_T, \quad \phi_T = 11.5^\circ \text{ lagging}$$

(c)

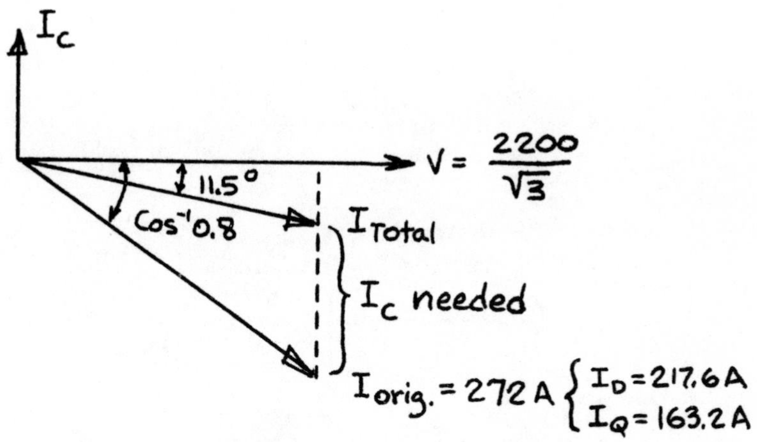

$$I_{Total} = \frac{I_D}{\cos 11.5^\circ} = \frac{217.6}{0.98} = 222.06$$

$$I_{Total_Q} = 222.06 \sin 11.5^\circ = 44.3 \text{ A}$$

$$\therefore I_C = I_{O_Q} - I_{T_Q} = 163.2 - 44.3 = 118.9 \text{ A}$$

$$\therefore \text{KVAR} = \frac{118.9 \times 2200}{1000\sqrt{3}} = 151 \text{ KVAR/Phase}$$

PROBLEM 5 **MEASUREMENT.** (Here problems may be from a bridge circuit using an a.c. source and converting to d.c. for a D'Arsonval meter
- or perhaps temperature measurement;
and, usually some form of wave analysis.)

A temperature-control system is being used in a chemical process. The temperature is controlled by a heater (and is assumed to be uniform throughout).

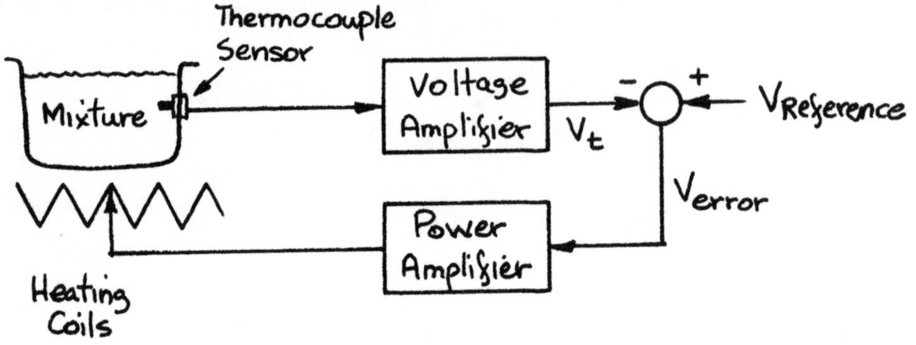

Assume the heater raises the mixture 10°C per K.W. of heating power and the power amplifier has a gain of 500 KW/volt.

The thermocouple sensor produces 0.003 m.v./°C and the voltage amplifier gain is 1000.

REQUIRED:

(a) Find the value of V_R if the mixture temperature should be 90°C. You may assume the ambient temperature is 25°C.

(b) Assume the power amplifier produces a full wave rectified voltage (60 hz) of 10 volts peak for it's output per 0.1 volt d.c. on it's input. Determine the necessary heating coil resistance so that the heating coil-power amplifier combination may be rated at 500 KW/volt.

SOLUTION

(a) Heater power required $= \dfrac{90° - 25°}{10°/KW} = 7.5$ KW

$$\therefore \text{amplifier input, } V_e = \frac{7.5 \text{ KW}}{500 \text{ KW/volt}} = 15 \times 10^{-3} \text{ volts}$$

at 90°C thermocouple produces $(90°)(0.003) = 0.27$ m.v.

$$\text{or } V_t = (1000)(0.27 \times 10^{-3}) = 0.270 \text{ volts}$$

Thus

$$V_R = V_t + V_e = 0.270 + 0.015 = 0.285 \text{ volts}$$

(b) 1 V. E_{in}(dc) produces 100 volts (peak) in output or 70.7 effective rms volts.

And, since the voltage "works into" a pure resistance heating coil,

$$P = E^2/R$$

Therefore, from 70.7 volts we need to produce 500 KW

$$500{,}000 = \frac{E^2}{R} = \frac{(70.7)^2}{R}$$

$$R = \frac{5000}{500{,}000} = 0.01\ \Omega$$

PROBLEM 6

Refer to the 10-part multiple choice problem in Chapter 7.

PROBLEM 7 **ELECTRONICS** (One might expect a problem on transmission lines, use of the Smith chart, field and waves, or a full communication systems problem. The actual problem presented here is Problem 8 from the Electronics chapter.

In the modulating circuit sketched below, the modulating signal (see spectrum sketch) is limited to angular frequencies where

$$\omega_{m_1} < \omega < \omega_{m_2} \lll \omega_c.$$

where ω_m = modulating frequency

ω_c = carrier frequency

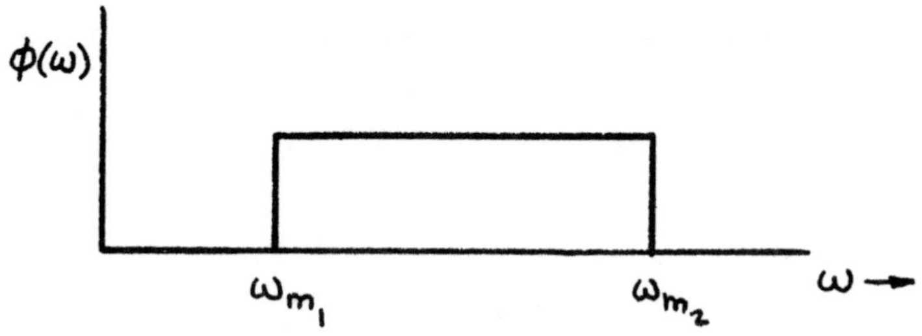

Spectrum of Modulating Signal

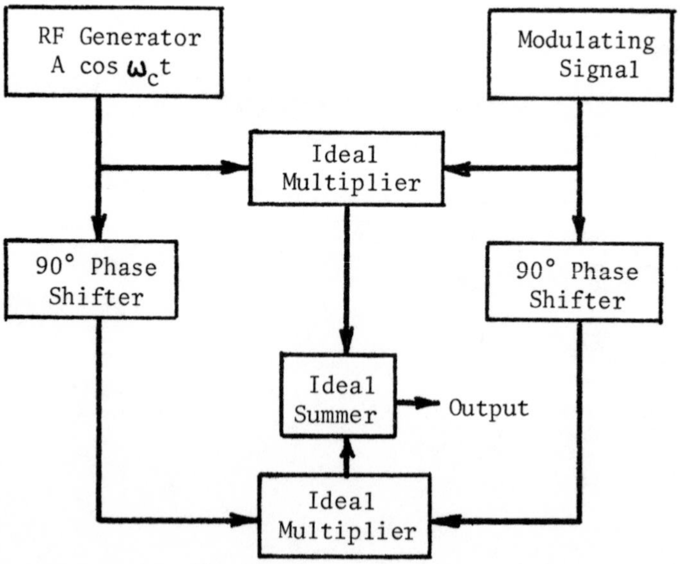

REQUIRED:

(a) Sketch the spectrum of the output.

(b) What is this type of modulated signal called?

(c) Sketch another circuit which would produce the same output spectrum.

Refer to Chapter 6 (Electronics)
Problem 8 for the solution.

PROBLEM 8 ELECTRONICS (Another frequently asked circuits problem is one involving filter theory, either constant "K", "M" derived, or similar to the following.)

Consider the following filter driven by a current source and the output goes to an infinite impedance load.
[Note: C_L is only in the circuit for Part (b).]

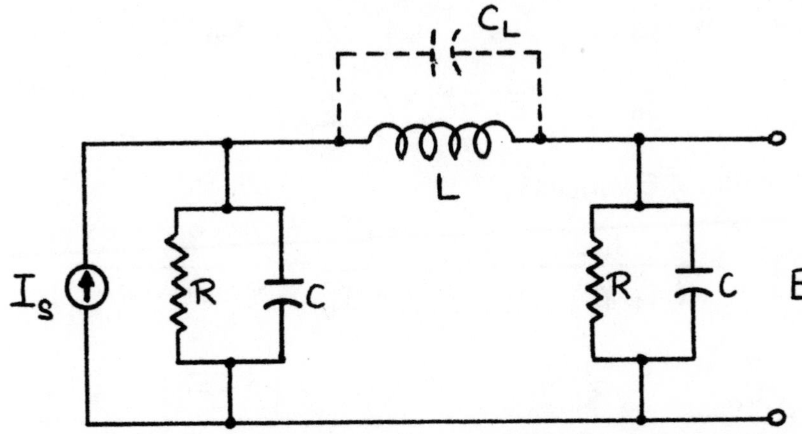

REQUIRED:

(a) Find the transfer function relating the output voltage to the input current, E_o/I_s.
From your solution describe the behavior of the circuit (i.e., high pass, low pass, maximally flat or what? Why?).

(b) If a capacitor, C_L, is placed across L, describe the behavior of the filter in terms of any specific important frequency.
Let $L = 0.01$ h, $C_L = C = 0.01\ \mu f$, $R = 100\ \Omega$

SOLUTION

(a)

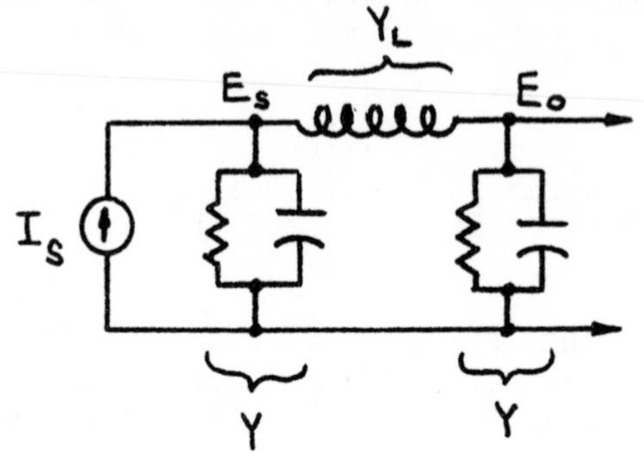

$$Y = \frac{1}{R} + SC = G + SC \qquad Y_L = \frac{1}{SL}$$

Σi's at E_s: $\quad I_s = E_s Y + (E_s - E_o) Y_L$

at E_o: $\quad (E_s - E_o) Y_L = E_o Y$

Eliminating E_s gives:

$$E_o = \frac{-Y_L I_s}{(Y+Y_L)^2 - Y_L}, \qquad \frac{E_o}{I_s} = \frac{Y_L}{Y^2 + 2YY_L}$$

$$\therefore H = \frac{E_o}{I_s} = \frac{\frac{1}{SL}}{(G+SC)^2 + 2(G+SC)\frac{1}{SL}} = \frac{\left(\frac{1}{C^2L}\right)}{(S+\frac{G}{C})\left[S^2 + \frac{G}{C}S + \frac{2}{LC}\right]}$$

Since the denominator is a cubic, two of the roots may be complex and one must be real. The pole locations turn out to be on a semi-circle and can be adjusted to 60° apart (the maximum separation the complex poles may have without peaks appearing in the output); this would yield a <u>maximally flat</u> low-pass filter design.

s-plane

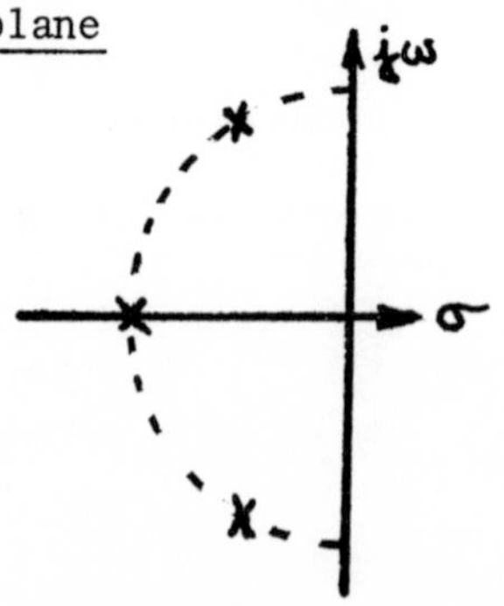

(b)

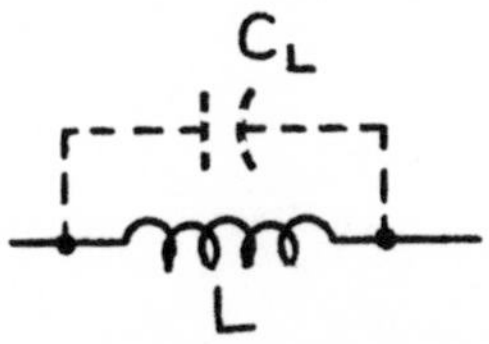

$$Y_L' = \frac{1}{SL} + SC_L$$

$$\therefore \frac{E_o}{I_s} = H = \frac{SC_L + \frac{1}{SL}}{(G+SC)\left[G+SC+2SC_L+\frac{2}{SL}\right]} \qquad \text{Let } C_x = C+2C_L$$

$$= \left(\frac{C_L}{CC_x}\right)\left[\frac{\left(S^2+\frac{1}{LC_L}\right)}{\left(S+\frac{G}{C}\right)\left(S^2+\frac{G}{C_x}S+\frac{2}{LC_x}\right)}\right]$$

The form of the denominator is the same as previously found (except that one can't get a maximally flat design); however, the numerator gives two complex zeros on the imaginary axis:

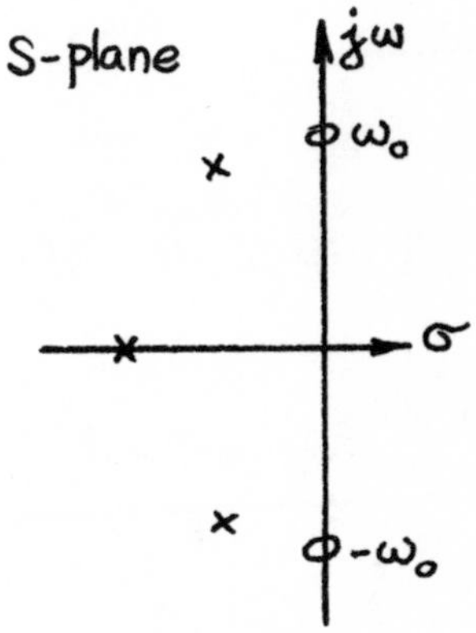

$$\omega_o = \sqrt{\frac{1}{LC_L}}$$

$$= \sqrt{\frac{1}{(10^{-2})(10^{-8})}}$$

$$= 10^5 \text{ rad/sec} \equiv 16 \text{ Khz}$$

Thus the 16 Khz signal will not pass.

PROBLEM 9 ECONOMICS.

There is one 10-part multiple choice Economics problem in the examination. An example of the problem is in Chapter 9.

PROBLEM 10 MACHINES.

(This area seems "wide open" as questions can be transformer connections, such as 3-phase zig-zag voltage ratios and loading; transformer efficiencies requiring the use of equivalent circuits from open circuit - short circuit data; complete d.c. motor analysis involving starting torques and efficiency calculations; a.c. machines and loading.)

A 12,500 K.V.A., 6600 volt, 3600 r.p.m., 60 hz, three-phase, Y connected alternator has magnetization and short circuit characteristic curves as shown on the next page.

REQUIRED: Determine the percentage voltage regulation for a 0.707 lagging power factor. Let the a.c. armature resistance be 0.5 ohms and make (and state) any reasonable assumption necessary for your solution.

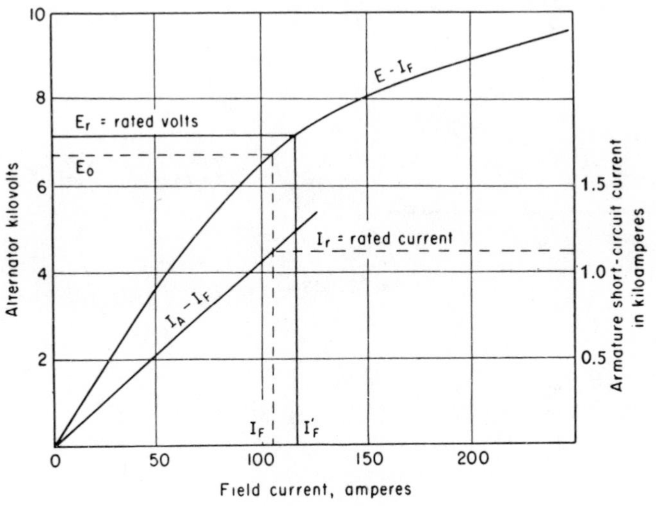

SOLUTION

$$\text{Rated } I = \frac{12{,}500{,}000}{6600\sqrt{3}} = 1090 \text{ A}$$

Field current necessary to give short circuit line current, from graph:

$$I_f \text{ (d.c.)} = 105 \text{ A}$$

Terminal voltage (per phase) from graph:

$$\frac{6300}{\sqrt{3}} = 3630 \text{ volts}$$

Terminal rated voltage (per phase)

$$\frac{6600}{\sqrt{3}} = 3800 \text{ volts}$$

(Here, one could find the synchronous reactance or impedance by taking these operating values and defining

$$X_s \approx \frac{3630 \text{ V}}{1090 \text{ A}},$$

however, since only one condition is asked for, the synchronous voltage drop is the 3630 volts.)

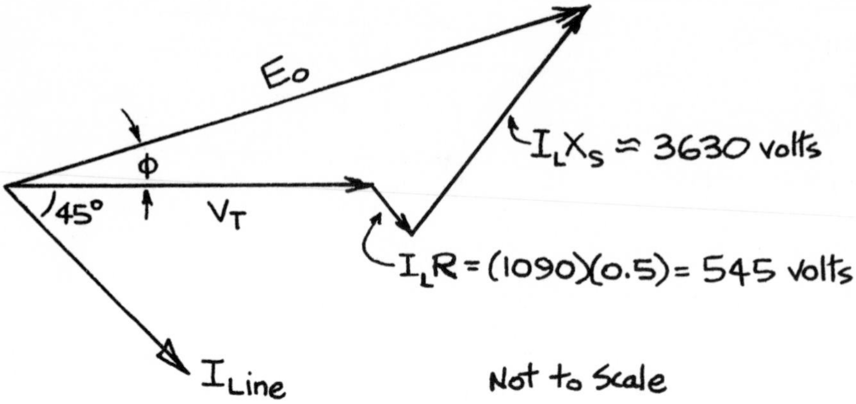

$$E_o = V_T + R_A I_L + j X_S I_L$$

$$= 3800 + 545(\cos 45° - j\sin 45°) + 3630(\cos 45° + j\sin 45°)$$

$$= 6752.4 + j\,2181.6 = 7096 \angle\phi \text{ volts}$$

$$\therefore \% \text{ V.R.} = \frac{E_o - V_T}{V_T}(100) = \frac{7096 - 3800}{3800}(100)$$

$$= 86.7\,\%$$

Note that the calculated no-load voltage is

$$7096\sqrt{3} = 12{,}290 \text{ volts}$$

which is well beyond the range of available field current; thus the saturated limit of no-load voltage would be between 9,000 and 10,000 volts. Thus the saturated value of synchronous reactance would be less than the value used here.

PROBLEM 11 DIGITAL LOGIC

Two binary coded digits, A and B, are set into the toggles (flip-flops) shown symbolically in the accompanying diagram. Each digit is composed of two bits, "1"s and "2"s, and can, therefore, have any value from zero through three.

The outputs at the top of each toggle have two possible voltage levels, -6.0 volts and approximately 0.0 volts. Each toggle set to the "1" state (or ON) has a "1" side output high (approximately 0.0 volts) and the "0" side output low (approximately -6.0 volts).

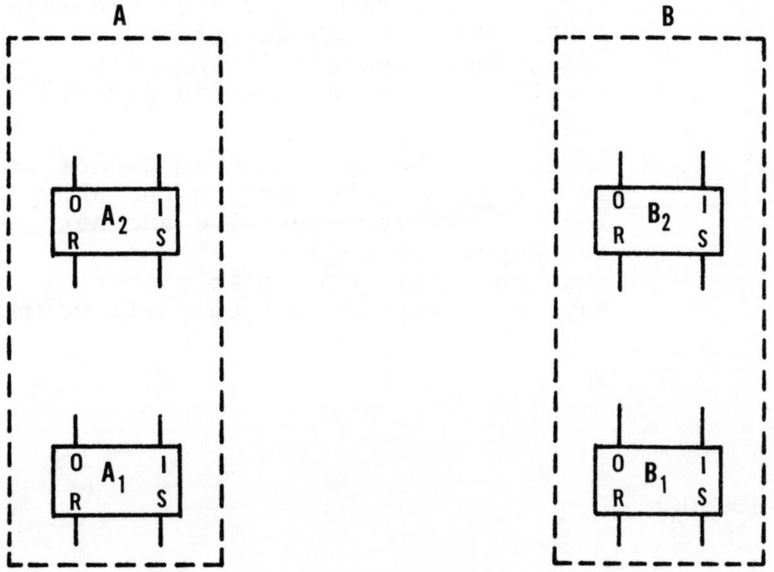

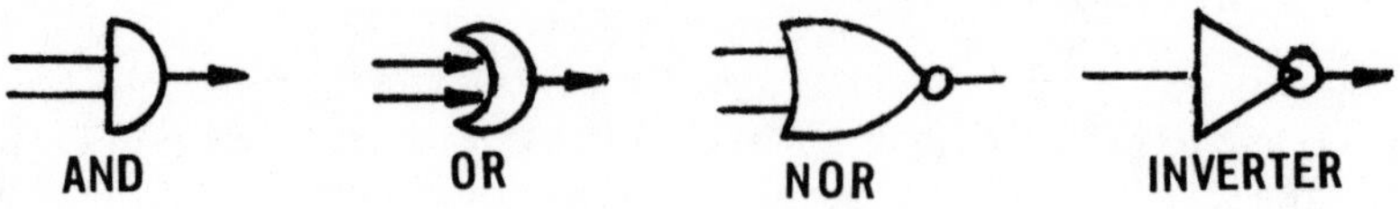

(1) <u>Wt. 6</u> Using the symbols given, show the logical gating and connections necessary to compare the digits (A and B) and to produce a high level (approx. 0.0 volts) output when, and only when, A is greater than B. Any usable type of logic may be assumed, such as diode logic, resistor-transistor logic, etc., or any combination of methods. Minimize the gates and components used as much as possible. Use the logical symbols shown for "AND", "OR", "NOR" gates or "INVERTERS". Or, you may use other, or additional, symbols if they are labelled and also shown in part (2), below.

(2) <u>Wt. 4</u> Sketch a schematic diagram typical of each <u>type</u> of gate used in the above logic. Assume a positive and negative supply voltage and ground. Show any gate returns to either supply or ground, but omit all values.

SOLUTION

The following truth table expresses the wording of the problem:

	A		B		
	A_2	A_1	B_2	B_1	f
0	0	0	0	0	0
1	0	1	0	1	0
2	0	0	1	0	0
3	0	0	1	1	0
4	0	1	0	0	1
5	0	1	0	1	0
6	0	1	1	0	0
7	0	1	1	1	0
8	1	0	0	0	1
9	1	0	0	1	1
10	1	0	1	0	0
11	1	0	1	1	0
12	1	1	0	0	1
13	1	1	0	1	1
14	1	1	1	0	1
15	1	1	1	1	0

The output "f" is to have a value - must be a "1" - if and only if the digit "A" is larger than the digit "B". The function "f" expressed in Minterm Canonical form would be (from the truth table):

$$f = M_4 + M_8 + M_9 + M_{12} + M_{13} + M_{14}$$

$$= \bar{A}_2 A_1 \bar{B}_2 \bar{B}_1 + A_2 \bar{A}_1 \bar{B}_2 \bar{B}_1 + A_2 \bar{A}_1 \bar{B}_2 B_1 + A_2 A_1 \bar{B}_2 \bar{B}_1$$

$$+ A_2 A_1 \bar{B}_2 B_1 + A_2 A_1 B_2 \bar{B}_1$$

If we choose to work with this type of expression, then the minimal form of the function can be found by a variety of techniques. The laws of Boolean algebra may be applied to obtain some reduction, a Karnaugh map or a Veitch diagram may be used to obtain the minimal form.

However, a better approach, especially for a more difficult problem, would be as follows (the Minterms correspond to the vertices of a N dimensioned cube, the covering of the cube gives the minimal form of the function):

f = T

4    $4/12 \rightarrow (4,3) \rightarrow A_1B_2B_1$

8
12    $8/9, 12/13 \rightarrow (8,2) \rightarrow A_2B_2$

$14/12 \rightarrow (12,1) \rightarrow A_2A_1B_1$

13
14

$$f_{min} = A_1\bar{B}_2\bar{B}_1 + A_2\bar{B}_2 + A_2A_1\bar{B}_1 \qquad (1)$$

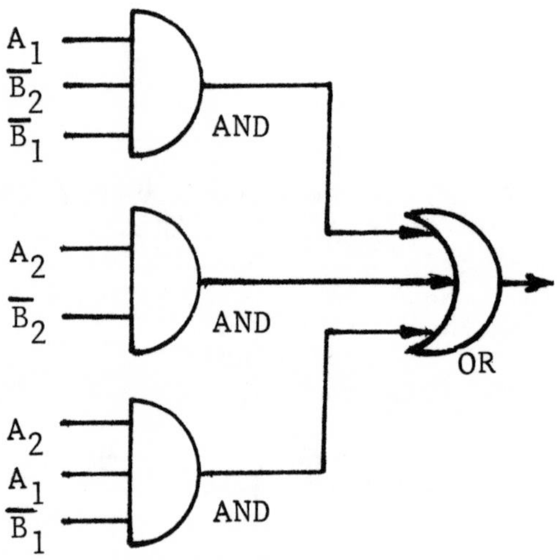

A second answer would be obtained by factoring the function as follows:

$$f = A_1\bar{B}_1\left[\bar{B}_2 + A_2\right] + A_2\bar{B}_2 \qquad (2)$$

This would add another stage of signal reduction and delay.

(2) Working with expression (1) of the first part of this problem, the AND gates would be designed as follows:

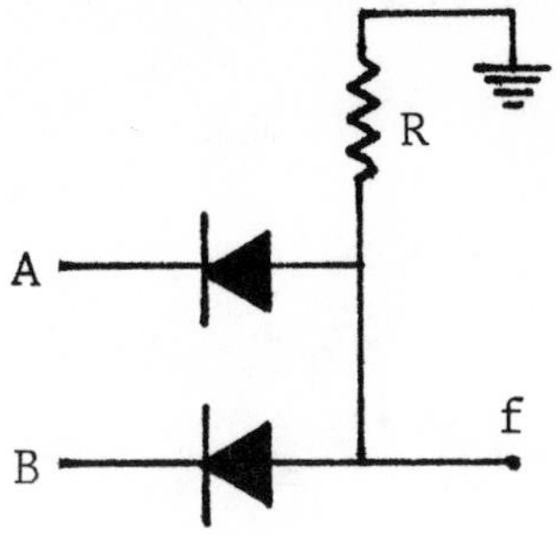

A	B	f
-6	-6	-6
-6	0	-6
0	-6	-6
0	0	0

With the above circuit, there would be an output if, and only if, both input "A" and input "B" were at the 0 voltage level.

The OR gate would be designed as follows:

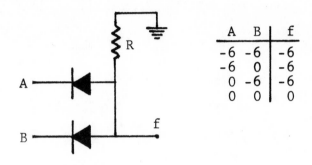

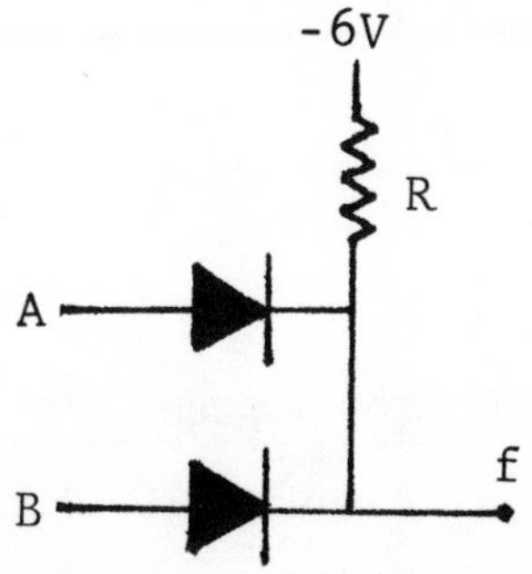

This circuit would have an output if either one or both inputs were at a zero voltage level.

PROBLEM 12 POWER

Resolve the system $E_A = 1500 \angle 30°$, $E_B = 1800 \angle -70°$, $E_C = 2000 \angle 170°$ into its symmetrical components.

SOLUTION

$$E_{A1} = \frac{1}{3}(1500 \angle 30° + 1800 \angle -70° + 120° + 2000 \angle 170° + 240°)$$

$$= \frac{1}{3}(1500 \times 0.866 + j1500 \times 0.5 + 1800 \times 0.642 + j1800 \times 0.766$$

$$+ 2000 \times 0.642 + j2000 \times 0.766)$$

$$= \frac{1}{3}(1300 + j750 + 1156 + j1379 + 1284 + j1532)$$

$$= \frac{1}{3}(3740 + j3661) = 1246 + j1220 = 1740 \angle 44°$$

$$E_{B1} = 1740 \angle 44° + 240° = 1740 \angle 284°$$

$$E_{C1} = 1740 \angle 44° + 120° = 1740 \angle 164°$$

$$E_{A2} = \frac{1}{3}(1500 \angle 30° + 1800 \angle -70 + 240° + 2000 \angle 170 + 120°$$

$$= \frac{1}{3}(1500 \times 0.866 + j1500 \times 0.5 - 1800 \times 0.985 + 1800 \times 0.174$$

$$+ 2000 \times 0.342 - j2000 \times 0.94)$$

$$= \frac{1}{3}(1300 + j750 - 1773 + j313 + 684 - j1880)$$

$$= \frac{1}{3}(211 - j817) = 70 - j272 = 281 \angle 284° = 281 \angle -76°$$

$$E_{B2} = 281 \angle -76° + 120° = 281 \angle 44°$$

$$E_{C2} = 281 \angle -76° + 240° = 281 \angle 164°$$

$$E_{A0} = \frac{1}{3}(1500 \angle 30° + 1800 \angle -70° + 2000 \angle 170°)$$

$$= \frac{1}{3}(1500 \times 0.866 + j1500 \times 0.5 + 1800 \times 0.342 - j1800 \times 0.94$$

$$- 2000 \times 0.985 + 2000 \times 0.174$$

$$= \frac{1}{3}(1300 + j750 + 616 - j1692 - 1970 + j348)$$

$$= \frac{1}{3}(-54 - j594) = 18 - j198 = 199 \angle 266°$$

$$E_{A0} = E_{B0} = E_{C0} = 199 \angle 266°$$

Reference: Elements of Power Systems Analysis by William D. Stevenson, Jr., Chapter 13.